CHAPMAN & HALL/CRC
Monographs and Surveys in
Pure and Applied Mathematics   128

HYPERBOLIC

CONSERVATION LAWS

AND THE COMPENSATED

COMPACTNESS METHOD

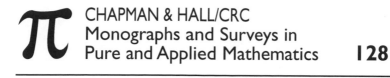

CHAPMAN & HALL/CRC
Monographs and Surveys in
Pure and Applied Mathematics   128

HYPERBOLIC

CONSERVATION LAWS

AND THE COMPENSATED

COMPACTNESS METHOD

YUNGUANG LU

CRC Press
Taylor & Francis Group
Boca Raton London New York

CRC Press is an imprint of the
Taylor & Francis Group, an **informa** business

A CHAPMAN & HALL BOOK

First published 2003 by Chapman & Hall

Published 2019 by CRC Press
Taylor & Francis Group
6000 Broken Sound Parkway NW, Suite 300
Boca Raton, FL 33487-2742

© 2003 by Taylor & Francis Group, LLC
CRC Press is an imprint of Taylor & Francis Group, an Informa business

First issued in paperback 2019

No claim to original U.S. Government works

ISBN-13: 978-0-367-45473-9 (pbk)
ISBN-13: 978-1-58488-238-1 (hbk)

Visit the Taylor & Francis Web site at
http://www.taylorandfrancis.com

and the CRC Press Web site at
http://www.crcpress.com

Library of Congress Card Number 2002073731

Library of Congress Cataloging-in-Publication Data

Lu, Yunguang.
Hyperbolic conservation laws and the compensated compactness method / Yunguang Lu.
p. cm. — (Monographs and surveys in pure and applied mathematics)
Includes bibliographical references and index.
ISBN 1-58488-238-7
1. Conservation laws (Mathematics) 2. Differential equations, Hyperbolic. I. Title. II.
Chapman & Hall/CRC monographs and surveys in pure and applied mathematics.

QA377 .H965 2003
515′.353—dc21   2002073731

In memory of my father, Cai-qin Lu

Contents

7 Le Roux System   71

8 System of Polytropic Gas Dynamics   85

9 Two Special Systems of Euler Equations   121

10 General Euler Equations of Compressible Fluid Flow   137

11 Extended Systems of Elasticity   147

Preface

I planned to write this book when I visited Professor Alan Jeffrey in Newcastle, England, in 1992. Since then, I have given seminars, courses and lectures on the applications of the compensated compactness method to hyperbolic conservation laws in many different universities. Among these are Stanford University, USA (1994), Heidelberg University, Germany (1995), International School for Advanced Studies (SISSA), Italy (1996), Federal University of Rio de Janeiro, Brazil (1998), National University of Colombia, Colombia (2000) and University of Science and Technology of China, China (2001).

This book is a result of a one-year course for graduate students in applied mathematics, but it can also be a textbook for undergraduate students in their last year. The students should be familiar with the basic contents of introductory courses such as functional analysis, measure theory, Sobolev space, shock waves theory and so on.

I want to thank Professor Alan Jeffrey for his continuing support and encouragement. Without his kind help, this book would never have been written.

I must also thank Mr. Ding-hao Li, my mathematics teacher in middle school, whose excellent character has always been a great influence in my life.

Thanks also go to my mother and all the members of my family, for their interest and sense of pride in every achievement in my career and who are my power resources to do mathematics.

Finally, I would like to thank Dr. Ben-jin Xuan, my former student, who helped me graph the figures and resolve all the technical problems in the Latex file during the last three months of my typing the manuscript. After I finished the process, he carefully read all the pages and proposed many valuable suggestions.

Chapter 1

Preliminary

Systems of hyperbolic conservation laws are very important mathematical models for a variety of physical phenomena that appear in traffic flow, theory of elasticity, gas dynamics, fluid dynamics and so on. In general, the classical solution of the Cauchy problem for nonlinear hyperbolic conservation laws exists only locally in time even if the initial data are small and smooth. This means that shock waves always appear in the solution for a suitable large time. Since the solution is discontinuous and does not satisfy the given partial differential equations in the classical sense, we have to study the generalized solutions, or functions which satisfy the equations in the sense of distributions.

We consider the quasi-linear systems of the form

$$u_t + f(u)_x = 0, \quad (x,t) \in R \times R^+, \tag{1.0.1}$$

where $u = (u_1, u_2, ..., u_n)^T \in R^n, n \geq 1$ is the unknown vector function standing for the density of physical quantities, $f(u) = (f_1(u), ..., f_n(u))^T$ is a given vector function denoting the conservative term. These equations are commonly called conservation laws. Let us suppose for the moment, that u is a classical solution of (1.0.1) with the initial data

$$u(x,0) = u_0(x). \tag{1.0.2}$$

Let C_0^1 be the class of C^1 function ϕ which vanishes outside of a compact subset. We multiply (1.0.1) by ϕ and integrate by parts over $t > 0$, to get

$$\iint_{t>0} (u\phi_t + f(u)\phi_x)dxdt + \int_{t=0} u_0\phi dx = 0. \tag{1.0.3}$$

Definition 1.0.1 *An $L^p, 1 < p \leq \infty$, bounded function $u(x,t)$ is called a weak solution of the initial-value problem (1.0.1) with L^p bounded initial data u_0, provided that (1.0.3) holds for all $\phi \in C_0^1(R \times R^+)$.*

An important aspect of the theory of nonlinear system of conservation laws is the question of existence of solutions to these equations. It helps to answer the question if the modelling of the natural phenomena at hand has been done correctly, and if the problem is well posed.

To get a global weak solution or a generalized solution for given hyperbolic conservation laws, a standard method is to add a small parabolic perturbation term to the right-hand side of (1.0.1):

$$u_t + f(u)_x = \varepsilon u_{xx}, \tag{1.0.4}$$

where $\varepsilon > 0$ is a constant.

We may first get a sequence of solutions $\{u^\varepsilon\}$ of the Cauchy problem (1.0.4),(1.0.2) for any fixed ε by the following general theorem for parabolic equations:

Theorem 1.0.2 *(1) For any fixed $\varepsilon > 0$, the Cauchy problem (1.0.4) with the bounded measurable initial data (1.0.2) always has a local smooth solution $u^\varepsilon(x,t) \in C^\infty(R \times (0,\tau))$ for a small time τ, which depends only on the L^∞ norm of the initial data $u_0(x)$.*

(2) If the solution u^ε has an a priori L^∞ estimate $|u^\varepsilon(\cdot,t)|_{L^\infty} \leq M(\varepsilon,T)$ for any $t \in [0,T]$, then the solution exists on $R \times [0,T]$.

(3) The solution u^ε satisfies:

$$\lim_{|x|\to\infty} u^\varepsilon = 0, \quad if \lim_{|x|\to\infty} u_0(x) = 0.$$

(4) Particularly, if one of the equations in system (1.0.4) is in the form

$$w_t + (wg(u))_x = \varepsilon w_{xx}, \tag{1.0.5}$$

where $g(u)$ is a continuous function of $u \in R^n$, then

$$w^\varepsilon \geq c(t, c_0, \varepsilon) > 0, \quad if \quad w_0(x) \geq c_0 > 0, \tag{1.0.6}$$

where c_0 is a positive constant and $c(t, c_0, \varepsilon)$ could tend to zero as the time t tends to infinity or ε tends to zero.

Proof. The local existence result in *(1)* can be easily obtained by applying the contraction mapping principle to an integral representation for a solution, following the standard theory of semilinear parabolic systems.

Whenever we have an *a priori* L^∞ estimate of the local solution, it is clear that the local time τ can be extended to T step by step since the step time depends only on the L^∞ norm.

The process to get the local solution clearly shows the behavior of the solution in *(3)*.

The details about the proofs of *(1)*-*(3)* in Theorem 1.0.2 can be seen in [LSU, Sm]. The following is the unpublished proof of (1.0.6) by Bereux and Sainsaulieu (cf. [Lu9, Pe]).

We rewrite Equation (1.0.5) as follows:

$$v_t + g(u)v_x + g(u)_x = \varepsilon(v_{xx} + v_x^2), \tag{1.0.7}$$

where $v = \log w$. Then

$$v_t = \varepsilon v_{xx} + \varepsilon(v_x - \frac{g(u)}{2\varepsilon})^2 - g(u)_x - \frac{g^2(u)}{4\varepsilon}. \tag{1.0.8}$$

The solution v of (1.0.8) with initial data $v_0(x) = \log(w_0(x))$ can be represented by a Green function $G(x - y, t) = \frac{1}{\sqrt{\pi ct}} exp(-\frac{(x-y)^2}{4\varepsilon t})$:

$$v = \int_{-\infty}^{\infty} G(x - y, t)v_0(y)dy$$
$$+ \int_0^t \int_{-\infty}^{\infty} \left(\varepsilon(v_x - \frac{g(u)}{2\varepsilon})^2 - \frac{g^2(u)}{4\varepsilon} - g(u)_x \right) G(x - y, t - s)dyds. \tag{1.0.9}$$

Since

$$\int_{-\infty}^{\infty} G(x - y, t)dy = 1, \qquad \int_{-\infty}^{\infty} |G_y(x - y, t)|dy \le \frac{M}{\sqrt{\varepsilon t}},$$

it follows from (1.0.9) that

$$
\begin{aligned}
v \;\geq\; & \int_{-\infty}^{\infty} G(x-y,t)v_0(y)dy \\
& + \int_0^t \int_{-\infty}^{\infty} \left(-\frac{g^2(u)}{4\varepsilon} - g(u)_x\right) G(x-y,t-s)dyds \\
= \; & \int_{-\infty}^{\infty} G(x-y,t)v_0(y)dy \\
& + \int_0^t \int_{-\infty}^{\infty} \left(g(u)G_y(x-y,t-s) - \frac{g^2(u)}{4\varepsilon} G(x-y,t-s) \right) dyds \\
\geq \; & \log c_0 - \frac{Mt}{\varepsilon} - \frac{M_1 t^{\frac{1}{2}}}{\varepsilon^{\frac{1}{2}}} \geq -C(t,c_0,\varepsilon) > -\infty.
\end{aligned}
$$

$$(1.0.10)$$

Thus w^ε has a positive lower bound $c(t, c_0, \varepsilon)$ for any fixed ε and $t < \infty$. ∎

The solution obtained in Theorem 1.0.2 is called viscosity solution. After we have the sequence of viscosity solutions $\{u^\varepsilon\}$, $\varepsilon > 0$, if we furthermore suppose that $\{u^\varepsilon\}$ are uniformly bounded in $L^p (1 < p \leq \infty)$ space with respect to the parameter ε, then there exists a subsequence (still labelled) $\{u^\varepsilon\}$ such that

$$u^\varepsilon(x,t) \rightharpoonup u(x,t), \quad \text{weakly in } L^p, \qquad (1.0.11)$$

and also a subsequence $\{f(u^\varepsilon)\}$ such that

$$f(u^\varepsilon(x,t)) \rightharpoonup l(x,t), \quad \text{weakly} \qquad (1.0.12)$$

under suitable growth conditions on $f(u)$. If

$$l(x,t) = f(u(x,t)), \quad a.e., \qquad (1.0.13)$$

then clearly $u(x,t)$ is a weak solution of system (1.0.1) with the initial data (1.0.2) by letting ε tend to zero in (1.0.4).

How could we obtain the weak continuity (1.0.13) of the nonlinear flux function $f(u)$ with respect to the sequence of viscosity solutions $\{u^\varepsilon\}$? The theory of compensated compactness is just to answer this question.

Why is this theory called Compensated Compactness? Roughly speaking, this term comes from the following fact:

If a sequence of functions satisfies

$$w^\varepsilon(x,t) \rightharpoonup w(x,t) \qquad (1.0.14)$$

with either

$$(w^\varepsilon)^2 + (w^\varepsilon)^3 \rightharpoonup w^2 + w^3 \text{ or } (w^\varepsilon)^2 - (w^\varepsilon)^3 \rightharpoonup w^2 - w^3 \qquad (1.0.15)$$

weakly as ε tends to zero, in general, $w^\varepsilon(x,t)$ is not compact. However, it is clear that any one weak compactness in (1.0.15) can compensate for another to make the compactness of w^ε. In fact, if we add them together, we get

$$(w^\varepsilon)^2 \rightharpoonup w^2 \qquad (1.0.16)$$

weakly as ε tends to zero, which combining with (1.0.14) implies the compactness of w^ε.

In this book, our goal is to introduce some applications of the method of compensated compactness to the scalar conservation law as well as some special systems of two or three equations. Moreover, applications to some physical systems with a relaxation perturbation parameter are also considered.

The arrangement of this book is as follows:

In Chapter 2, we introduce some elemental theorems in the theory of compensated compactness. Section 2.1 is about the weak continuity theorems of 2×2 determinants, and the proofs come from [Ta]. Section 2.2 is about the Young measure representation theorems of weak limits and we use the proofs in [Lin]. Section 2.3 is about the Murat compact embedding theorems. In this part, we introduce two theorems. The proof of Theorem 2.3.2 is the same as that given in [DCL1] and the proof of Theorem 2.3.4 is copied from the French paper by Murat [Mu]. It is necessary to point out that Theorem 2.3.4 is independent of this book and the readers could pass over it without considering the details. We collect it here because it was used in some research papers (cf. [CLL, JPP]).

In Chapter 3, we consider the Cauchy problem of the scalar equation with L^∞ and $L^p(1 < p < \infty)$ initial data, respectively. In this part, a simplified proof (cf. [CL1, Lu1]) without using the Young measure is given.

In the first part of Chapter 4, we introduce some basic definitions of systems of two equations, such as the strict hyperbolicity, genuine nonlinearity, linear degeneration, Riemann invariants, entropy-entropy flux pair and so on (cf. [La2, La3, Sm]).

In the second part, a framework to obtain L^∞ estimates of viscosity solutions for systems of two equations, called the Invariant Region Theory from [CCS], is introduced.

In Chapter 5, we consider a special symmetric system of two equations ([Ch3]). This system is very similar to the scalar equation because one characteristic field is always linearly degenerate, although the other field is genuinely nonlinear. This system is of interest because along the genuinely nonlinear characteristic field, the compactness of viscosity solutions is obtained in L^∞ space without any more regular condition. However, along the linearly degenerate characteristic field, some more regular conditions, such as BV estimates, must be added to ensure the strong compactness of the sequence of viscosity solutions.

In Chapter 6, we consider a system of two equations with quadratic flux. This system is nonstrictly hyperbolic at the original point, one characteristic field is linearly degenerate on the positive half axis of u, while the other field is linearly degenerate on the negative half axis of u. Its entropy equation is the same as that of the system of polytropic gas dynamics with the adiabatic exponent $\gamma = 2$. The main difficulty in studying this system by the compensated compactness is that the entropy-entropy flux pairs are singular at the original point. Through a careful construction of exact solutions of the classical Fuchsian equation, we obtain the explicit entropy-entropy flux pair of Lax type for this system. Then the necessary estimates for the major terms of these entropies follow from the analysis of solutions of the Fuchsian equation (cf. [Lu4]).

In Chapter 7, we extend the method given in Chapter 6 to the Le Roux system, which is also nonstrictly hyperbolic at the original point, but the entropy equation is the same as that of $\gamma = 5/3$ for the polytropic gas. This system is of interest because it is a typical system of Temple type, whose characteristic fields are both straight lines. The proof in this chapter is from [LMR].

In Chapter 8, we consider the most typical hyperbolic conservation system of two equations, the so-called system of the polytropic gas dynamics (or γ-law). For the case of $\gamma > 3$, our proof is copied from the paper [LPT]. For the case of $1 < \gamma \leq 3$, using only four pairs of weak entropy-entropy flux, we give a short proof by assuming that the solution is away from vacuum and small (cf. [CL2]).

In Chapter 9, the methods in Chapters 6 and 7 are again extended to study two special systems of one-dimensional Euler equations, which are nonstrictly hyperbolic on the vacuum line $\rho = 0$. For smooth solutions, they are equivalent to the systems of polytropic gas dynamics with the adiabatic exponents $3 < \gamma < \infty$ and $\gamma = \infty$, respectively. Our proofs in this chapter come from [Lu2] and [Lu8].

In Chapter 10, we consider the general Euler equations of one-dimensional, compressible fluid flow. This more general system is again nonstrictly hyperbolic on the vacuum line $\rho = 0$. To study this system by using the compensated compactness, one basic difficulty is how to construct entropy-entropy flux pairs and obtain the necessary estimates on these entropies. Since the method to construct entropy-entropy flux pairs of Lax type (cf. [La1]) to strictly hyperbolic systems does not work here, in this chapter we extend DiPerna's method to nonstrictly hyperbolic systems. We introduce a special form of Lax entropy, in which the progression terms are functions of a single variable. The necessary estimates for the major terms are obtained by the singular perturbation theory of the ordinary differential equations of second order. The proof in this chapter comes from [Lu6].

In Chapter 11, we extend the method given in Chapter 10 to study some extended systems of elasticity in L^∞ space. The proof is also from [Lu6].

In Chapter 12, some important results about $L^p, 1 < p < \infty$, weak solutions for the system of elasticity are introduced, which include a compactness framework of artificial viscosity solutions to this system by Lin [Lin] and a compactness framework of physical viscosity by Shearer [Sh]. An application of the latter compactness framework by Shearer on the system of adiabatic gas flow through porous media is also considered (cf. [LK1]).

However, to avoid knotty mathematical formulas, we choose not to provide the proofs of these two compactness frameworks in this book, although they are very important and form a basis on relaxation problems of hyperbolic systems of three equations in Chapter 16.

From Chapter 13 to Chapter 16, we introduce some applications of the compensated compactness on the relaxation problems.

In Chapter 13, a general description of the relaxation singular problem is introduced.

In Chapter 14, singular limits of stiff relaxation and dominant diffusion for general 2×2 nonlinear systems of conservation laws are considered. These include the L^∞ solutions of the system of elasticity, the system of isentropic fluid dynamics in Eulerian coordinates and the extended models of traffic flows; the $L^p, 1 < p < \infty$, solutions for some physical models, without L^∞ bounded estimates, such as the system of isentropic fluid dynamics in Lagrangian coordinates, and the models of traffic flows in different states. All proofs in this chapter can be found in [Lu9].

In Chapter 15, a framework for the singular limits of stiff relaxation for general 2×2 hyperbolic conservation systems (not necessary strictly hyperbolic) is introduced (cf. [CLL], [Lu10]).

An application of this framework on a nonstrictly hyperbolic system, the so-called system of extended traffic flow, is also obtained. The proof is from [Lu10].

In Chapter 16, singular limits of stiff relaxation and dominant diffusion for the general 3×3 system of chemical reaction are considered (cf. [Lu12]). The pure relaxation limit (without viscosity) for a special case of this chemical reaction system is also introduced (cf. [LK1, LK2, Tz]).

Chapter 2

Theory of Compensated Compactness

As a theory, the compensated compactness is a large subject. However, until now, all the applications on hyperbolic conservation laws were related to the theorems given in this chapter.

2.1 Weak Continuity of a 2×2 Determinant

Theorem 2.1.1 *Let $H^{-1}(\Omega)$ be the dual of $H_0^1(\Omega)$ and $\Omega \subset R^N$ be an open, bounded set. Suppose*

(H_1) $u^\varepsilon = (u_1^\varepsilon, u_2^\varepsilon, \cdots, u_p^\varepsilon) \rightharpoonup u = (u_1, u_2, \cdots, u_p)$ *weakly in $L^2(\Omega)$,*

(H_2) $\displaystyle\sum_{j=1}^{p} \sum_{k=1}^{N} a_{ijk} \frac{\partial u_j^\varepsilon}{\partial x_k}$ *are compact in the strong topology of H_{loc}^{-1},*

where $i = 1, 2, ...q$.

Then if $Q = Q(\lambda), \lambda \in R^p$ is quadratic and satisfies $Q(\lambda) \geq 0$ for all $\lambda \in \wedge$, where

$$\wedge = \{\lambda \in R^p : \exists \xi \in R^N - \{0\}, \ s.t. \ \sum_{j=1}^{p} \sum_{k=1}^{N} a_{ijk} \lambda_j \xi_k = 0 \quad i = 1, 2, ...q\};$$

$$(2.1.1)$$

if $Q(u^\varepsilon) \rightharpoonup l$ in the sense of distributions (l may be a measure), then

$$l \geq Q(u) \quad \text{in the sense of distributions.} \qquad (2.1.2)$$

Proof. *Step 1:* We make a translation

$$v_i^\varepsilon = u_i^\varepsilon - u_i.$$

Then (H_1) and (H_2) imply that

(1) $\quad v_j^\varepsilon \rightharpoonup 0$ weakly in $L^2(\Omega)$ for $j = 1, 2, ...p$,

(2) $\quad \sum_{j,k} a_{ijk} \dfrac{\partial v_j^\varepsilon}{\partial x_k}$ are compact in the strong topology of H_{loc}^{-1} for

$i = 1, 2, ...q.$

Since Q is quadratic, there exists a bilinear form $q(a, b)$ such that

$$Q(a) = q(a, a).$$

Therefore

$$Q(v^\varepsilon) = Q(u^\varepsilon - u) = Q(u^\varepsilon) - 2q(u^\varepsilon, u) + Q(u).$$

But

$$Q(u^\varepsilon) \rightharpoonup l, \quad q(u^\varepsilon, u) \rightharpoonup q(u, u) \quad \text{weakly,}$$

the second statement holding because $q(a, b)$ is linear in a for fixed b. Therefore since $q(u, u) = Q(u)$,

$$Q(v^\varepsilon) \rightharpoonup l - Q(u), \quad \text{weakly.}$$

Step 2: Next we perform a localization as follows. We let

$$w^\varepsilon = \phi v^\varepsilon \quad \text{with } \phi \in C_0^\infty(\Omega).$$

Then

$$\operatorname{supp} w^\varepsilon \subset \text{compact set of } R^N \tag{2.1.3}$$

and

$$w^\varepsilon \rightharpoonup 0 \text{ weakly in } L^2(\Omega). \tag{2.1.4}$$

Moreover

$$\sum_{j,k} a_{ijk} \frac{\partial w_j^\varepsilon}{\partial x_k} = \phi \sum_{j,k} a_{ijk} \frac{\partial v_j^\varepsilon}{\partial x_k} + \sum_{j,k} a_{ijk} v_j^\varepsilon \frac{\partial \phi}{\partial x_k},$$

and the first term in the right-hand side belongs to a compact set of $H^{-1}(\Omega)$. Therefore

$$\sum_{j,k} a_{ijk} \frac{\partial w_j^\varepsilon}{\partial x_k} \in \text{ compact set of } H^{-1}(\Omega).$$

Hence extracting a possible subsequence we have

$$\lim_{\varepsilon \to 0} \sum_{j,k} a_{ijk} \frac{\partial w_j^\varepsilon}{\partial x_k} = 0 \text{ strongly in } H^{-1}(\Omega) \qquad (2.1.5)$$

and

$$Q(w^\varepsilon) \rightharpoonup \phi^2(l - Q(u)) \qquad \text{weakly.}$$

Step 3: To prove $l - Q(u) \geq 0$, it is enough to show that

$$\lim_{\varepsilon \to 0} \int Q(w^\varepsilon) dx \geq 0, \qquad (2.1.6)$$

since this will prove that

$$\lim_{\varepsilon \to 0} \int Q(\phi v^\varepsilon) dx \geq 0 \iff \int \phi^2(l - Q(u)) dx \geq 0$$

for all $\phi \in C_0^\infty(\Omega)$, implying that $l - Q(u) \geq 0$.

Step 4: Let us define the Fourier transformation of w_j^ε as

$$\widehat{w}_j^\varepsilon = F(w_j^\varepsilon) \int_{R^N} w_j^\varepsilon(x) e^{-2\pi i(\xi \cdot x)} dx. \qquad (2.1.7)$$

The Plancherel formula gives

$$\int_{R^N} v(x) \bar{w}(x) dx = \int_{R^N} \widehat{v}(\xi) \bar{\widehat{w}}(\xi) d\xi, \qquad (2.1.8)$$

where v and $w \in L^2(R^N)$ are complex valued functions.

We extend Q from R^p to $\mathbb{C}^p$ into an Hermitian form. Recall that Q is quadratic and takes the form

$$Q(\lambda) = \sum_{j,k} q_{jk} \lambda_j \lambda_k$$

with real coefficients $q_{jk} = q_{kj}$. We define

$$\widetilde{Q}(\lambda) = \sum_{j,k} q_{jk} \lambda_j \overline{\lambda_k}.$$

Hence we have

$$\operatorname{Re}\left(\widetilde{Q}(\lambda)\right) \geq 0 \qquad \text{if } \lambda \in \wedge + i \wedge. \tag{2.1.9}$$

In fact, if $\lambda = \lambda_1 + i\lambda_2$ with $\lambda_1, \lambda_2 \in \wedge$, then

$$\widetilde{Q}(\lambda) = (Q(\lambda_1) + Q(\lambda_2)) + i(q(\lambda_1, \lambda_2) + q(\lambda_2, \lambda_1)),$$

where q was defined by $Q(a) = q(a, a)$, and therefore

$$\operatorname{Re}\left(\widetilde{Q}(\lambda)\right) = Q(\lambda_1) + Q(\lambda_2) \geq 0.$$

By the Plancherel formula again,

$$\int_{R^N} Q(w^\varepsilon)dx = \int_{R^N} \widetilde{Q}(\widehat{w}^\varepsilon)d\xi = \int_{R^N} \operatorname{Re} \widetilde{Q}(\widehat{w}^\varepsilon)d\xi.$$

So (2.1.6) is equivalent to

$$\lim_{\varepsilon \to 0} \int_{R^N} \operatorname{Re} \widetilde{Q}(\widehat{w}^\varepsilon)d\xi \geq 0. \tag{2.1.10}$$

But, since $\operatorname{supp} w_j^\varepsilon \subset$ fixed compact set C of R^N, we have

$$\widehat{w}_j^\varepsilon(\xi) = \int_C w_j^\varepsilon(x)e^{-2\pi i(\xi \cdot x)}dx.$$

But $e^{-2\pi i(\xi \cdot x)} \in L^2(C)$ and since $w_j^\varepsilon \rightharpoonup 0$ weakly in $L^2(R^N)$, we deduce that

$$\lim_{\varepsilon \to 0} \widehat{w}_j^\varepsilon(\xi) = 0 \qquad \text{strongly for all } \xi, \quad \text{and } \|\widehat{w}_j^\varepsilon(\xi)\| \leq M. \tag{2.1.11}$$

Therefore

$$\lim_{\varepsilon \to 0} \widehat{w}_j^\varepsilon = 0 \text{ locally in } L^2(R^N). \tag{2.1.12}$$

Hence

$$\lim_{\varepsilon \to 0} \int_{|\xi| \leq r} \widetilde{Q}(\widehat{w}^\varepsilon)d\xi = 0 \tag{2.1.13}$$

for any fixed constant $r > 0$.

Using the Fourier transformation of (2.1.5) leads to

$$\lim_{\varepsilon \to 0} \frac{1}{1 + |\xi|} \sum_{j,k} a_{ijk} \widehat{w}_j^\varepsilon(\xi) \xi_k = 0 \text{ strongly in } L^2(R^N) \text{ for } i = 1, ..., q.$$

(2.1.14)

Using the following lemma, we can finish the proof of Theorem 2.1.1.

Lemma 2.1.2 *Suppose that* $Q(\lambda) \geq 0$ *for all* $\lambda \in \wedge$. *Then for all* $\alpha > 0$, *there exists a constant* C_α *such that*

$$Re \, \widetilde{Q}(\lambda) \geq -\alpha |\lambda|^2 - C_\alpha \sum_{i=1}^{q} \Big| \sum_{j,k} a_{ijk} \lambda_j \eta_k \Big|^2$$

(2.1.15)

for all $\lambda \in \mathbb{C}^p$, *and for all* $\eta \in R^N$ *with* $|\eta| = 1$.

In fact, it follows from (2.1.14) that

$$\lim_{\varepsilon \to 0} \frac{1}{|\xi|} \sum_{j,k} a_{ijk} \widehat{w}_j^\varepsilon(\xi) \xi_k = 0 \text{ strongly in } L^2(\{|\xi| \geq 1\}).$$

(2.1.16)

Using Lemma 2.1.2 and taking $\lambda = \widehat{w}^\varepsilon(\xi)$ and $\eta = \xi/|\xi|$, we have

$$Re \, \widetilde{Q}(\widehat{w}^\varepsilon(\xi)) \geq -\alpha |\widehat{w}^\varepsilon(\xi)|^2 - C_\alpha \sum_{i=1}^{q} \Big| \sum_{j,k} a_{ijk} \widehat{w}_j^\varepsilon(\xi) \xi_k/|\xi| \Big|^2.$$ (2.1.17)

Integrating this on $|\xi| \geq 1$, we get

$$\int_{|\xi| \geq 1} Re \, \widetilde{Q}(\widehat{w}^\varepsilon(\xi)) d\xi \geq -\alpha \int_{|\xi| \geq 1} |\widehat{w}^\varepsilon(\xi)|^2 d\xi$$

$$-C_\alpha \int_{|\xi| \geq 1} \sum_{i=1}^{q} \Big| \sum_{j,k} a_{ijk} \widehat{w}_j^\varepsilon(\xi) \xi_k/|\xi| \Big|^2 d\xi.$$

(2.1.18)

But by (2.1.11) and (2.1.16) we deduce that

$$\lim_{\varepsilon \to 0} \int_{|\xi| \geq 1} Re \, \widetilde{Q}(\widehat{w}^\varepsilon(\xi)) d\xi \geq -\alpha M$$

(2.1.19)

for all α as small as desired. Hence

$$\lim_{\varepsilon \to 0} \int_{|\xi| \geq 1} \operatorname{Re} \widetilde{Q}(\widehat{w}^\varepsilon(\xi)) d\xi \geq 0. \tag{2.1.20}$$

Combining (2.1.13) with (2.1.20) gives the proof of (2.1.10), and hence (2.1.6), which implies the proof of Theorem 2.1.1. ∎

Proof of Lemma 2.1.2. To prove Lemma 2.1.2, we proceed by contradiction. Suppose there exists an $\alpha > 0$ such that for all $C_\alpha = n$ there exist $\lambda^n \in \mathbb{C}^p$ with $|\lambda^n| = 1$ and $\eta^n \in R^N$ with $|\eta^n| = 1$ such that

$$\operatorname{Re} \widetilde{Q}(\lambda^n) < -\alpha |\lambda^n|^2 - n \sum_{i=1}^q \Big| \sum_{j,k} a_{ijk} \lambda_j^n \eta_k^n \Big|^2. \tag{2.1.21}$$

Extract convergent subsequences such that $\lambda^n \to \lambda^\infty$ and $\eta^n \to \eta^\infty$. Then

$$\sum_{i=1}^q \Big| \sum_{j,k} a_{ijk} \lambda_j^n \eta_k^n \Big|^2 \leq \frac{C}{n},$$

where C is a constant. Hence, passing to the limit, we deduce that

$$\sum_{j,k} a_{ijk} \lambda_j^\infty \eta_k^\infty = 0, \quad i = 1, \cdots, q.$$

Therefore

$$\lambda^\infty \in \wedge + i\wedge,$$

and hence

$$\operatorname{Re} \widetilde{Q}(\lambda^\infty) \geq 0.$$

But from (2.1.21) we have also

$$\operatorname{Re} \widetilde{Q}(\lambda^\infty) = \lim_{n \to \infty} \operatorname{Re} \widetilde{Q}(\lambda^n) \leq -\alpha < 0$$

and this is a contradiction. ∎

Corollary 2.1.3 *If Q is quadratic and satisfies $Q(\lambda) = 0$ for all $\lambda \in \wedge$, and if $\{u^\varepsilon\}$ satisfies (H_1) and (H_2), then $Q(u^\varepsilon) \to Q(u)$ in the sense of distributions.*

Proof. Extract a subsequence such that

$$Q(u^\varepsilon) \rightharpoonup l, \ l \text{ may be a measure.}$$

Applying Theorem 2.1.1 to Q and then to $-Q$, we obtain $l = Q(u)$ and hence the proof of Corollary 2.1.3 since this is true for all subsequences. ∎

Theorem 2.1.4 *(weak continuity theorem of a 2×2 determinant) Let $\Omega \subset R \times R^+$ be a bounded open set and $u^\varepsilon : \Omega \to R^4$ be measurable functions satisfying*

$$u^\varepsilon \rightharpoonup u, \ in \ (L^2(\Omega))^4 \tag{2.1.22}$$

and

$$\frac{\partial u_1^\varepsilon}{\partial t} + \frac{\partial u_2^\varepsilon}{\partial x}, \ \frac{\partial u_3^\varepsilon}{\partial t} + \frac{\partial u_4^\varepsilon}{\partial x} \ are \ compact \ in \ H_{loc}^{-1}(\Omega). \tag{2.1.23}$$

Then there exists a subsequence (still labelled) $\{u^\varepsilon\}$ such that

$$\begin{vmatrix} u_1^\varepsilon & u_2^\varepsilon \\ u_3^\varepsilon & u_4^\varepsilon \end{vmatrix} \rightharpoonup \begin{vmatrix} u_1 & u_2 \\ u_3 & u_4 \end{vmatrix} \ in \ the \ sense \ of \ distributions. \tag{2.1.24}$$

Proof. From the conditions in (2.1.23), the set $\wedge$ in Theorem 2.1.1 consists of all points $\lambda \in L^4$ satisfying

$$\lambda_1 \xi_1 + \lambda_2 \xi_2 = 0, \quad \lambda_3 \xi_1 + \lambda_4 \xi_2 = 0$$

for any $\xi \in R \times R^+ - \{0\}$, or equivalently

$$\wedge = \{\lambda \in L^4 : \lambda_1 \lambda_4 - \lambda_2 \lambda_3 = 0\}. \tag{2.1.25}$$

Let $Q(\lambda)$ in Theorem 2.1.1 be $\lambda_1 \lambda_4 - \lambda_2 \lambda_3$. Then clearly $Q(\lambda) = 0$ for all $\lambda \in \wedge$. So (2.1.24) follows from Corollary 2.1.3. This completes the proof of Theorem 2.1.4. ∎

2.2 Young Measure Representation Theorems

Theorem 2.2.1 *Let $\Omega \subset R^N$ be measurable. Suppose that $\{u^n(x)\}$ is a sequence of measurable functions from Ω to R^S. Then there exists a subsequence $\{u^{n_k}\}$ of $\{u^n(x)\}$ and a family of positive measures $\nu_x \in M(R^S)$, depending measurably on $x \in \Omega$, such that for any $f \in C_0(R^S)$,*

$$w^\star - \lim f(u^{n_k}) = < f(\lambda), \nu_x(\lambda) > \int_{R^S} f(\lambda) d\nu_x(\lambda), \qquad (2.2.1)$$

where $M(R^S)$ is the dual space of $C_0(R^S)$ and $C_0(R^S)$ is the space of continuous functions which tend to zero at infinity, and $w^\star - \lim$ denotes the weak-star limit in the L^∞ space. Furthermore, if the range of $u^n(x)$ is contained in $G \subset R^S$, then so is the support of ν_x; if the L^∞ norm of $u^n(x)$ is uniformly bounded or G is compact in R^S, then ν_x are probability measures, i.e., the mass of ν_x is one.

Proof. Let $E = \{f^m\}$ be a dense set in $C_0(R^S)$. Then $\{f^1(u^n)\}$ is bounded on Ω, and hence there exist a subsequence $\{u^{n^1_k}\}$ of $\{u^n\}$ and $(\alpha(f^1)(x) \in L^\infty(\Omega)$ such that

$$w^\star - \lim f^1(u^{n^1_k}) = \alpha(f^1)(x). \qquad (2.2.2)$$

Furthermore, $\{f^2(u^{n^1_k})\}$ is also bounded on Ω, and hence there exist a subsequence $\{u^{n^2_k}\}$ of $\{u^{n^1_k}\}$ and $\alpha(f^2)(x) \in L^\infty(\Omega)$ such that

$$w^\star - \lim f^2(u^{n^2_k}) = \alpha(f^2)(x). \qquad (2.2.3)$$

Proceeding in this way we obtain a subsequence $\{u^{n^m_k}\}, \alpha(f^m)(x)$ such that

(i) $\{u^{n^1_k}\} \supset \{u^{n^2_k}\} \supset \{u^{n^3_k}\} \supset \cdots$, and

(ii) for each fixed m, $w^\star - \lim f^m(u^{n^m_k}) = \alpha(f^m)(x)$.

We let $\{u^{n_k}\} = \{u^{n^k_k}\}$, the diagonal sequence. Then from (ii) we get that for each fixed m,

$$w^\star - \lim f^m(u^{n_k}) = \alpha(f^m)(x). \qquad (2.2.4)$$

For each $f^m \in E$, we define a bounded functional $I(f^m)$ on $L^1(\Omega)$ by

$$< I(f^m), \psi > = \int_\Omega \psi \alpha(f^m) dx = \lim_{k \to \infty} \int_\Omega \psi f^m(u^{n_k}) dx, \quad \forall \psi \in L^1(\Omega).$$
$$(2.2.5)$$

Then for any given $f \in C_0(R^S)$, suppose that $f = \lim_{l \to \infty} f^l$ in $C_0(R^S)$, where $\{f^l\} \subset E$. We now prove that the following limit exists, and hence denote it by $I(f)$, namely,

$$< I(f), \psi >= \lim_{k \to \infty} \int_\Omega \psi f(u^{n_k}) dx, \quad \forall \psi \in L^1(\Omega). \qquad (2.2.6)$$

In fact, for any $u^{n_{k_1}}, u^{n_{k_2}}$, we notice that

$$\left| \int_\Omega \psi[f(u^{n_{k_1}}) - f(u^{n_{k_2}})] dx \right|$$

$$\leq \left| \int_\Omega \psi[f(u^{n_{k_1}}) - f^l(u^{n_{k_1}})] dx \right| + \left| \int_\Omega \psi[f(u^{n_{k_2}}) - f^l(u^{n_{k_2}})] dx \right|$$

$$+ \left| \int_\Omega \psi[f^l(u^{n_{k_1}}) - f^l(u^{n_{k_2}})] dx \right|$$

$$\leq 2\|f - f^l\|_{C_0}\|\psi\|_1 + \left| \int_\Omega \psi[f^l(u^{n_{k_1}}) - f^l(u^{n_{k_2}})] dx \right|. \qquad (2.2.7)$$

We first choose l large enough such that the first term on the right-hand side of (2.2.7) is small. Then by (2.2.4) the second term on the right-hand side of (2.2.7) can be small whenever n_{k_1} and n_{k_2} are large enough. Hence we prove that $\{\int_\Omega \psi f(u^{n_k}) dx\}$ is a Cauchy sequence for any fixed $\psi \in L^\infty(\Omega)$, and so we have proved (2.2.6). Consequently, we obtain

$$| < I(f), \psi > | \leq \|f\|_{C_0}\|\psi\|_1, \quad \forall \psi \in L^1(\Omega). \qquad (2.2.8)$$

We notice, by (2.2.8), that $I(f)$ is a bounded functional on $L^1(\Omega)$, and hence by the Riesz representation theorem there exists $\alpha(f)(x) \in L^\infty(\Omega)$ such that

$$< I(f), \psi >= \int_\Omega \alpha(f)\psi dx, \quad \forall \psi \in L^1(\Omega). \qquad (2.2.9)$$

We also have

$$\alpha(f_1 + f_2) = \alpha(f_1) + \alpha(f_2) \quad \forall f_i \in C_0(R^S), \quad i = 1, 2,$$

$$\alpha(kf) = k\alpha(f) \quad \forall f \in C_0(R^S), \quad k \in R.$$

At this moment we suppose, without loss of generality, that every point $x \in \Omega$ is a Lebesgue point of each function $\alpha(f)$. Then for any fixed $x_0 \in \Omega$, we set

$$\psi(x) = (\text{meas } B_r(x_0))^{-1}\chi_{B_r(x_0)},$$

where $B_r(x_0)$ is the ball centered at x_0 with radius r, and $\chi_{B_r(x_0)}$ is the characteristic function of $B_r(x_0)$. By (2.2.8) and (2.2.9) we get

$$|(\text{meas } B_r(x_0))^{-1}\int_{B_r(x_0)} \alpha(f)dx| \leq \|f\|_{C_0}.$$

We now pass to the limit as $r \to 0$ to obtain $|\alpha(f)(x_0)| \leq \|f\|_{C_0}$. Combining this with the fact that $\alpha(f)$ is linear with respect to f we have that $\alpha(f)(x_0)$ is a bounded functional on $C_0(R^S)$. Therefore, applying the Riesz representation theorem we have a $\nu_{x_0} \in M(R^S)$ such that

$$\alpha(f)(x_0) = < f(\lambda), \nu_{x_0} > = \int_{R^S} f(\lambda)d\nu_{x_0}.$$

Since x_0 is arbitrary we get

$$< I(f), \psi > = \int_\Omega \psi < f(\lambda), \nu_x > dx \quad \forall \psi \in L^1(\Omega),$$

where $\nu_x \in M(R^S)$ for almost all $x \in \Omega$. So we have proved (2.2.1).

Furthermore, we notice that for any positive $f \in C_0(R^S)$,

$$\alpha(f)(x) = < f(\lambda), \nu_x > \geq 0 \text{ a.e. } x \in \Omega,$$

which implies that ν_x is positive for almost all $x \in \Omega$.

It is obvious from (2.2.1) that the support of ν is the same as the range of $u^n(x)$. In fact, we can choose any function $f \in C_0(R^S)$ satisfying supp $f \subset R^S - \{G\}$. Then clearly

$$0 = w^\star - \lim f(u^{n_k}) = \int_{R^S} f(\lambda)d\nu_x(\lambda),$$

which implies that the support of ν is contained in G.

Finally, if the L^∞ norm of $u^n(x)$ is uniformly bounded or G is compact in R^S, then we choose $f \equiv 1$ and get $\int_G d\nu_x(\lambda) = 1$ from (2.2.1). Thus the mass of ν_x is one, which shows that ν_x are probability measures. This completes the proof of Theorem 2.2.1. ∎

Corollary 2.2.2 *Suppose that* $\{u^n(x)\}$ *is bounded in* $L^p_{loc}(R^N; R^S)$, *where* $1 \le p < \infty$. *Then there exist a subsequence* $\{u^{n_k}\}$ *of* $\{u^n\}$ *and a family of positive measures* $\nu_x \in M(R^S), x \in R^N$, *such that for any bounded set* $A \subset R^N$,

$$w - \lim f(u^{n_k}) = < f(\lambda), \nu_x > \quad in \ L^1(A), \tag{2.2.10}$$

whenever $f \in C(R^S)$ *satisfies*

$$\lim_{|\lambda| \to \infty} \frac{f(\lambda)}{|\lambda|^p} = 0. \tag{2.2.11}$$

Proof. Without loss of generality we assume that $f \ge 0$. Then we define $f^m \in C_0(R^S)$ by $f^m = \theta^m f$, where $\theta^m \in C_0(R^S)$ is defined by

$$\theta^m(\lambda) = \begin{cases} 1 & \text{for } |\lambda| \le m, \\ 1 + m - |\lambda| & \text{for } m \le |\lambda| \le m+1, \\ 0 & \text{for } |\lambda| \ge m+1. \end{cases}$$

We claim that for each $\phi \in L^\infty(A)$,

$$\lim_{m \to \infty} \int_A \phi f^m(u^n) dx = \int_A \phi f(u^n) dx \tag{2.2.12}$$

uniformly in n. Indeed,

$$\left| \int_A \phi[f^m(u^n) - f(u^n)] dx \right| \le \|\phi\|_{L^\infty(A)} \int_{\{x \in A: |u^n| \ge m\}} f(u^n) dx$$

$$\le \|\phi\|_{L^\infty(A)} \|u\|_{L^p(A)} \max_{|\lambda| \ge m} \{\frac{f(\lambda)}{|\lambda|^p}\}, \tag{2.2.13}$$

which tends to 0 uniformly in n as $m \to \infty$.

On the other hand, by Theorem 2.2.1 there exist a subsequence $\{u^{n_k}\}$ of $\{u^n\}$ and a family of positive measures $\nu_x \in M(R^S)$ such that for each m,

$$\lim_{n \to \infty} \int_A \phi f^m(u^{n_k}) dx = \int_A \phi < f^m, \nu_x > dx \quad \forall \phi \in L^\infty(A). \tag{2.2.14}$$

Furthermore, from the monotone convergence theorem we get

$$\lim_{n \to \infty} \int_A \phi < f^m, \nu_x > dx = \int_A \phi < f, \nu_x > dx. \qquad (2.2.15)$$

Combining (2.2.12), (2.2.14) and (2.2.15), we get (2.2.10) and complete the proof of Corollary 2.2.2. ∎

Theorem 2.2.3 *Let $1 < p < \infty$ and suppose that the support of ν_x in Corollary 2.2.2 is a point for a.e. $x \in R^N$. Then there exists a subsequence $\{u^{n_k}\}$ of $\{u^n\}$ which converges strongly to u in $L^q_{loc}(R^N)$ for $1 \le q < p$. Furthermore, $\nu_x = \delta_{u(x)}$ for a.e. $x \in R^N$.*

Proof. From Corollary 2.2.2 and the assumption that the support of ν_x is a point, we have that $\nu_x = \delta_{v(x)}$ for a.e. $x \in R^N$ and for some function $v(x)$. Since $1 < p < \infty$, we can let $f(\lambda) = \lambda_i$ for $i = 1, 2, \cdots, S$, respectively in Corollary 2.2.2, to obtain

$$u_i^n = g(u^n) \rightharpoonup < f(\lambda), \nu_x >=< \lambda_i, \nu_x >= v_i, \qquad (2.2.16)$$

i.e., u^n converges weakly to v and hence $v = u$ because of the uniqueness of weak limit.

Now define $f_i(\lambda) = |\lambda_i|^q$ for any $1 \le q < p$. Again using Corollary 2.2.2, we get for any $\phi \in C_0(R^N)$ supported on compact set $K \in R^N$,

$$\int_K \phi(x)|u_i^n|^q dx \; \to \; \int_K \phi(x) \int_{R^S} f_i(\lambda) d\nu_x(\lambda) dx$$
$$= \int_K \phi(x)|u_i|^q dx \qquad (2.2.17)$$

as n tends to ∞. Thus $u^n \to u$ strongly in $L^q_{loc}(R^N)$ for any $1 \le q < p$, and this completes the proof of Theorem 2.2.3. ∎

2.3 Embedding Theorems

In this section we shall establish a compactness embedding theorem which is a generalization of Murat Lemma [Mu] or just an interpolation inequality (cf. [Tr]).

We first introduce the following lemma:

Lemma 2.3.1 *Let $\Omega \subset R^N$ be bounded and open, and $f \in W^{-1,p}(\Omega)$ for $1 < p < \infty$, supp $f \subset\subset \Omega$. Let u be the solution of equation*

$$\begin{cases} -\Delta u = f, & in \ \Omega, \\ u = 0, & on \ \partial\Omega. \end{cases} \qquad (2.3.1)$$

Then

$$\|u\|_{W^{1,p}(\Omega)} \leq C\|f\|_{W^{-1,p}(\Omega)}, \qquad (2.3.2)$$

where C is a positive constant independent of f.

This lemma can be proved by the L^p-theory of elliptic equations (cf. [Si]).

Theorem 2.3.2 *Let $\Omega \subset R^N$ be bounded and open, C be a compact set in $W_{loc}^{-1,q}(\Omega), B$ be a bounded set in $W_{loc}^{-1,r}(\Omega)$ for some constants q, p, r satisfying $1 < q \leq p < r < \infty$. Furthermore, let $D \subset \mathcal{D}(\Omega)$ such that $D \subset C \cap B$. Then there exists E, a compact set in $W_{loc}^{-1,p}(\Omega)$ such that $D \subset E$.*

Proof. For any set $D \subset C \cap B$, there exists a convergent subsequence $\{f_k\} \subset D$ such that

$$\|f_i\|_{W_{loc}^{-1,r}(\Omega)} \leq M, \quad \|f_i - f_j\|_{W_{loc}^{-1,q}(\Omega)} \to 0, \quad (i, j \to \infty), \qquad (2.3.3)$$

where M is independent of i.

Furthermore, for any $\Omega_1 \subset\subset \Omega$, we take a function $\phi \in C_0^\infty(\Omega)$ satisfying $\phi|_{\Omega_1} \equiv 1$ and define $\bar{f}_i \equiv \phi f_i$. Then $\bar{f}_i$ satisfies supp $\bar{f}_i \subset\subset \Omega$ and $\bar{f}_i|_{\Omega_1} = f_i|_{\Omega_1}$.

Now let u_i be solutions of equations

$$\begin{cases} -\Delta u_i = \bar{f}_i, & in \ \Omega, \\ u_i = 0, & on \ \partial\Omega. \end{cases} \qquad (2.3.4)$$

Then

$$\|u_i\|_{W^{1,q}(\Omega)} \leq C\|\bar{f}_i\|_{W^{-1,q}(\Omega)}, \quad \|u_i\|_{W^{1,r}(\Omega)} \leq C\|\bar{f}_i\|_{W^{-1,r}(\Omega)}. \qquad (2.3.5)$$

Hence from (2.3.3) and (2.3.5), we have

$$\begin{aligned} \|u_i - u_j\|_{W^{1,q}(\Omega)} &\leq C\|\bar{f}_i - \bar{f}_j\|_{W^{-1,q}(\Omega)} \\ &\leq C\|f_i - f_j\|_{W_{loc}^{-1,q}(\Omega)} \to 0, \quad (i, j \to \infty). \end{aligned} \qquad (2.3.6)$$

Substituting $u_i - u_j$ into the interpolation inequality

$$\|v\|_{W^{1,p}(\Omega)} \le \|v\|_{W^{1,r}(\Omega)}^{1-\alpha} \cdot \|v\|_{W^{1,q}(\Omega)}^{\alpha}, \tag{2.3.7}$$

for a constant $\alpha \in (0,1)$, we have

$$\begin{aligned}
\|u_i - u_j\|_{W^{1,p}(\Omega)} &\le \|u_i - u_j\|_{W^{1,r}(\Omega)}^{1-\alpha} \cdot \|u_i - u_j\|_{W_{loc}^{1,q}(\Omega)}^{\alpha} \\
&\le C_1 \|u_i - u_j\|_{W_{loc}^{1,q}(\Omega)}^{\alpha} \to 0, \quad (i,j \to \infty).
\end{aligned} \tag{2.3.8}$$

Therefore,

$$\begin{aligned}
&\|f_i - f_j\|_{W^{-1,p}(\Omega_1)} \\
&= \sup_{\|\phi\|_{W_0^{1,p'}(\Omega)} \le 1,\ \text{supp}\ \phi \subset \Omega_1} | < f_i - f_j, \phi > | \\
&\le \sup_{\|\phi\|_{W_0^{1,p'}(\Omega)} \le 1} | < \bar{f}_i - \bar{f}_j, \phi > | \|\bar{f}_i - \bar{f}_j\|_{W^{-1,p}(\Omega)} \\
\end{aligned} \tag{2.3.9}$$

$$\begin{aligned}
&= \|\triangle u_i - \triangle u_j\|_{W^{-1,p}(\Omega)} \\
&\le \|u_i - u_j\|_{W^{1,p}(\Omega_1)} \to 0, \quad (i,j \to \infty),
\end{aligned}$$

where p' is the constant satisfying

$$\frac{1}{p} + \frac{1}{p'} = 1.$$

This means that D is a compact set in $W_{loc}^{-1,p}(\Omega)$. So we get the proof of Theorem 2.3.2. ∎

Lemma 2.3.3 *Suppose that $\Omega \subset R^N$ is a bounded domain with Lipschitz boundary $\partial\Omega$. Then for any $\eta > 0$ small enough, $1 < p' < q' < +\infty$, there exists a function $\psi^\eta \in \mathcal{D}(\Omega)$, such that for any $\varphi \in W_0^{1,q'}(\Omega)$, there holds*

$$\|(1 - \psi^\eta)\varphi\|_{W_0^{1,p'}(\Omega)} \le \eta \|\varphi\|_{W_0^{1,q'}(\Omega)}. \tag{2.3.10}$$

Proof. *Step 1: Partition of unity.*

Since Ω is bounded and $\partial\Omega$ is Lipschitz, there exists an open covering $\beta_0, \beta_1, \cdots, \beta_m$ of $\bar{\Omega}$, such that

$$\beta_0 \cap \partial\Omega = \emptyset, \ \beta_j \cap \partial\Omega \ne \emptyset, \ (1 \le j \le m),$$

and for $1 \leq j \leq m$, there exist an open ball $B_j \subset R^N$ and a bijective Lipschitz mapping $T_j : \beta_j \to B_j$, such that $B_j \cap R_+^N = T_j(\beta_j \cap \Omega)$.

Let $\alpha_j = \beta_j \cap \Omega$, $A_j = B_j \cap R_+^N$, $(1 \leq j \leq m)$. Suppose $\Phi_0, \cdots, \Phi_m$ is partition of unity associated with covering $\beta_0, \beta_1, \cdots, \beta_m$, i.e.,

$$\Phi_i \in \mathcal{D}(\beta_i), \ \Phi_i \geq 0, \ (0 \leq i \leq m),$$

and

$$\sum_{i=0}^{m} \Phi_i = 1 \text{ on } \bar{\Omega}.$$

Step 2: Construction of a function $\bar{\psi}^\eta$ such that $\text{supp}\,\bar{\psi}^\eta \subset\subset \Omega$.

For any $j(1 \leq j \leq m)$ and $\eta > 0$ small enough, define a function θ_j^η on A_j (which depends only on x_N) as

$$\theta_j^\eta = \begin{cases} 0, & \text{if } 0 \leq x_N \leq \eta/2, \\ \dfrac{1}{\eta}(x_N - \dfrac{\eta}{2}), & \text{if } \eta/2 \leq x_N \leq \eta, \\ 1, & \text{if } \eta \leq x_N. \end{cases}$$

Let

$$\bar{\psi}^\eta = \Phi_0 + \sum_{j=1}^{m} (\theta_j^\eta \circ T_j)\Phi_j,$$

where $\circ$ denotes the composition of two mappings. Then function $\bar{\psi}^\eta$ is Lipschitz and $\text{supp}\,\bar{\psi}^\eta \subset\subset \Omega$.

Step 3: For any $\eta > 0$, there exist a function $\psi^\eta \in \mathcal{D}(\Omega)$ and a positive constant C_N which only depends on N, such that for any $\varphi \in W_0^{1,q'}(\Omega)$, $(1 < p' < q' < +\infty)$, there holds

$$\|(\psi^\eta - \bar{\psi}^\eta)\varphi\|_{W_0^{1,p'}(\Omega)} \leq C_N \eta \|\varphi\|_{W_0^{1,q'}(\Omega)}. \tag{2.3.11}$$

In fact, let $1 < p' < q' < +\infty$, define s as

$$\frac{1}{q'} + \frac{1}{s} = \frac{1}{p'},$$

hence $p' < s < +\infty$. For any $\eta > 0$, there exists a function $\psi^\eta \in \mathcal{D}(\Omega)$ such that

$$\|\psi^\eta - \bar{\psi}^\eta\|_{W_0^{1,s}(\Omega)} \leq \eta. \tag{2.3.12}$$

Then for any $\varphi \in W_0^{1,q'}(\Omega)$, there holds

$$\|(\psi^\eta - \bar{\psi}^\eta)\varphi\|_{W_0^{1,p'}(\Omega)}$$

$$\leq C_N \Big\{ \|(\psi^\eta - \bar{\psi}^\eta)\varphi\|_{L^{p'}(\Omega)} + \sum_{k=1}^N \|(\psi^\eta - \bar{\psi}^\eta)\tfrac{\partial \varphi}{\partial x_k}\|_{L^{p'}(\Omega)}$$

$$+ \sum_{k=1}^N \|\tfrac{\partial(\psi^\eta - \bar{\psi}^\eta)}{\partial x_k}\varphi\|_{L^{p'}(\Omega)} \Big\}.$$

From the Hölder inequality, there hold

$$\|(\psi^\eta - \bar{\psi}^\eta)\varphi\|_{L^{p'}(\Omega)} \leq \|\psi^\eta - \bar{\psi}^\eta\|_{L^s(\Omega)} \|\varphi\|_{L^{q'}(\Omega)},$$

$$\|(\psi^\eta - \bar{\psi}^\eta)\tfrac{\partial \varphi}{\partial x_k}\|_{L^{p'}(\Omega)} \leq \|\psi^\eta - \bar{\psi}^\eta\|_{L^s(\Omega)} \|\varphi\|_{W_0^{1,q'}(\Omega)}, \quad k = 1, \cdots, N$$

and

$$\|\tfrac{\partial(\psi^\eta - \bar{\psi}^\eta)}{\partial x_k}\varphi\|_{L^{p'}(\Omega)} \leq \|\psi^\eta - \bar{\psi}^\eta\|_{W_0^{1,s}(\Omega)} \|\varphi\|_{L^{q'}(\Omega)}, \quad k = 1, \cdots, N.$$

Thus, there holds

$$\|(\psi^\eta - \bar{\psi}^\eta)\varphi\|_{W_0^{1,p'}(\Omega)} \leq C_N \|\psi^\eta - \bar{\psi}^\eta\|_{W_0^{1,s}(\Omega)} \|\varphi\|_{W_0^{1,q'}(\Omega)}. \qquad (2.3.13)$$

(2.3.12) and (2.3.13) imply (2.3.11).

Step 4: For any $\eta > 0$, there exists a positive constant C_Ω, such that for any $\varphi \in W_0^{1,q'}(\Omega)$, there holds

$$\|(1 - \bar{\psi}^\eta)\varphi\|_{W_0^{1,p'}(\Omega)} \leq C_\Omega \eta^{1/s} \|\varphi\|_{W_0^{1,q'}(\Omega)}. \qquad (2.3.14)$$

Since $1 - \bar{\psi}^\eta = \sum_{j=1}^m (1 - \theta_j^\eta \circ T_j)\Phi_j$, it suffices to show that for any $1 \leq j \leq m$, there exists a constant $C > 0$ such that for any $\varphi \in W_0^{1,q'}(\Omega)$, there holds

$$\|(1 - \theta_j^\eta \circ T_j)\Phi_j\varphi\|_{W_0^{1,p'}} \leq C\eta^{1/s} \|\varphi\|_{W_0^{1,q'}(\Omega)}. \qquad (2.3.15)$$

For simplicity, we will drop the index j. For any $\varphi \in W_0^{1,q'}(\Omega)$, there holds

$$\|(1 - \theta^\eta \circ T)\Phi\varphi\|_{W_0^{1,p'}}$$

$$\leq C_N \Big\{ \|(1 - \theta^\eta \circ T)\Phi\varphi\|_{L^{p'}} + \sum_{k=1}^N \|(1 - \theta^\eta \circ T)\frac{\partial(\Phi\varphi)}{\partial x_k}\|_{L^{p'}}$$

$$+ \sum_{k=1}^N \|\frac{\partial(1-\theta^\eta \circ T)}{\partial x_k}\Phi\varphi\|_{L^{p'}} \Big\}.$$

$$(2.3.16)$$

Since mapping $T : \alpha = \beta \cap \Omega \to A = B \cap R_+^N$ is bi-Lipschitz, changing the variables of integration, there holds:

$$\|(1 - \theta^\eta \circ T)\Phi\varphi\|_{L^{p'}(\alpha)} = \Big\{ \int_\alpha |1 - \theta^\eta \circ T|^{p'}|\Phi\varphi|^{p'}\,dx \Big\}^{1/p'}$$

$$= \Big\{ \int_A |1 - \theta^\eta|^{p'}|(\Phi\varphi) \circ T^{-1}|^{p'}|\mathrm{Jac}(T^{-1})|\,dx \Big\}^{1/p'}$$

$$\leq \|1 - \theta^\eta\|_{L^s(A)}\|(\Phi\varphi) \circ T^{-1}\|_{L^{q'}(A)}\|\mathrm{Jac}(T^{-1})\|_{L^\infty(A)}^{1/p'},$$

where $\mathrm{Jac}(T^{-1})$ denotes the Jacobian of mapping T^{-1}.

Let $A^\eta = \{x \in A \mid 0 < x_N < \eta\}$. Then there holds

$$\|1 - \theta^\eta\|_{L^s(A)} \leq \|1\|_{L^s(A^\eta)} = (\mathrm{meas}\, A^\eta)^{1/s} \leq C_A \eta^{1/s}.$$

Thus, there holds

$$\|(1 - \theta^\eta \circ T)\Phi\varphi\|_{L^{p'}(\alpha)}$$

$$(2.3.17)$$

$$\leq C_A \eta^{1/s} C_T \|\Phi\varphi\|_{L^{q'}(\alpha)} \leq C\eta^{1/s}\|\varphi\|_{L^{q'}(\Omega)},$$

where C depends only on T and A.

Similarly, for $1 \leq k \leq N$, there holds

$$\|(1 - \theta^\eta \circ T)\frac{\partial(\Phi\varphi)}{\partial x_k}\|_{L^{p'}(\alpha)}$$

$$(2.3.18)$$

$$\leq C_A \eta^{1/s} C_T \|\frac{\partial(\Phi\varphi)}{\partial x_k}\|_{L^{q'}(\alpha)} \leq C\eta^{1/s}\|\varphi\|_{W_0^{1,q'}(\Omega)},$$

where C depends only on T, A and Φ.

Finally, since θ^η only depends on x_N, there holds

$$\frac{\partial(\theta^\eta \circ T)}{\partial x_N} = \frac{\partial\theta^\eta}{\partial x_N} \circ T \frac{\partial T_N}{\partial x_N},$$

and $\partial \theta^\eta / \partial x_N = 0$ outside of A^η. Then from the Hölder inequality and changing the variables of integration, there holds

$$\left\| \frac{\partial(\theta^\eta \circ T)}{\partial x_N} \Phi \varphi \right\|_{L^{p'}(\alpha)}$$

$$= \left\{ \int_{A^\eta} \left| \frac{\partial \theta^\eta}{\partial x_N} \left(\frac{\partial T_N}{\partial x_N} \right) \circ T(\Phi \varphi) \circ T \right|^{p'} |\mathrm{Jac}(T^{-1})| \, dx \right\}^{1/p'}$$

$$\leq \left\| \frac{\partial \theta^\eta}{\partial x_N} \right\|_{L^s(A)} \left\| \left(\frac{\partial T_N}{\partial x_N} \right) \circ T \right\|_{L^\infty(A)}$$

$$\cdot \left\| (\Phi \varphi) \circ T^{-1} \right\|_{L^{q'}(A^\eta)} \| \mathrm{Jac}(T^{-1}) \|_{L^\infty(A)}^{1/p'}.$$

From the definition of θ^η, there holds

$$\left\| \frac{\partial \theta^\eta}{\partial x_N} \right\|_{L^s(A)} \leq \frac{2}{\eta} (\mathrm{meas}\, A^\eta)^{1/s} \leq \frac{2}{\eta} C_A \eta^{1/s}.$$

Note that for any $g \in W_0^{1,q'}(A)$, there holds

$$\| g \|_{L^{q'}(A^\eta)} \leq \eta \left\| \frac{\partial g}{\partial x_N} \right\|_{L^{q'}(A^\eta)}.$$

Let $g = (\Phi \varphi) \circ T^{-1}$. Then there holds

$$\left\| (\Phi \varphi) \circ T^{-1} \right\|_{L^{q'}(A^\eta)}$$

$$\leq \eta \sum_{l=1}^{N} \left\| \frac{\partial(\Phi \varphi)}{\partial x_l} \circ T^{-1} \right\|_{L^{q'}(A)} \left\| \frac{\partial T_l^{-1}}{\partial x_N} \right\|_{L^\infty(A)} \sum_{l=1}^{N} \left\| \frac{\partial(\Phi \varphi)}{\partial x_l} \right\|_{L^{q'}(A)}$$

$$\leq \eta C_T \sum_{l=1}^{N} \left\| \frac{\partial(\Phi \varphi)}{\partial x_l} \right\|_{L^{q'}(\alpha)}.$$

Thus for $1 \leq k \leq N$, there holds

$$\left\| \frac{\partial(\theta^\eta \circ T)}{\partial x_N} \Phi \varphi \right\|_{L^{p'}(\alpha)} \leq \frac{2}{\eta} C_A \eta^{1/s} C_T \eta \sum_{l=1}^{N} \left\| \frac{\partial(\Phi \varphi)}{\partial x_l} \right\|_{L^{q'}(\alpha)}$$

$$(2.3.19)$$

$$\leq C \eta^{1/s} \| \varphi \|_{L^{q'}(\Omega)},$$

where C depends only on T, A and Φ.

Then (2.3.16), (2.3.17), (2.3.18) and (2.3.19) imply (2.3.15), which finishes the proof of Lemma 2.3.3. ∎

Theorem 2.3.4 *(Murat theorem) Suppose that $\Omega \subset R^N$ is an open domain, $1 < p < +\infty$, $\frac{1}{p} + \frac{1}{p'} = 1$. If $\{f^\varepsilon\}_{\varepsilon>0}$ satisfies:*

$$f^\varepsilon \rightharpoonup f^0 \text{ weakly in } W^{-1,p}(\Omega), \quad as \ \varepsilon \to 0, \tag{2.3.20}$$

$$f^\varepsilon \geq 0, \tag{2.3.21}$$

i.e., for any $\varphi \in \mathcal{D}(\Omega)$, $\varphi \geq 0$, there holds

$$< f^\varepsilon, \varphi > \leq 0, \tag{2.3.22}$$

where $< \cdot, \cdot >$ denotes the pairing between $W^{-1,p}$ and $W_0^{1,p'}$. Then

$$f^\varepsilon \to f^0 \text{ strongly in } W_{loc}^{-1,q}(\Omega), \quad as \ \varepsilon \to 0, \ \forall q < p. \tag{2.3.23}$$

If furthermore, Ω is bounded with Lipschitz boundary $\partial\Omega$, then

$$f^\varepsilon \to f^0 \text{ strongly in } W^{-1,q}(\Omega), \quad as \ \varepsilon \to 0, \ \forall q < p. \tag{2.3.24}$$

Proof. *Step 1:* For any $K \subset\subset \Omega$, there exists C_K such that for any $\varphi \in \mathcal{D}(\Omega)$, $\operatorname{supp}\varphi \subset K$, $\varepsilon > 0$, there holds

$$| < f^\varepsilon, \varphi > | \leq C_K \|\varphi\|_{L^\infty(\Omega)}. \tag{2.3.25}$$

In fact, suppose for any $K \subset\subset \Omega$, there exists a function Φ_K such that: $\Phi_K \in \mathcal{D}(\Omega)$, $\Phi_K \equiv 1$ on K and $\Phi_K \geq 0$ in Ω. Then for any $\varphi \in \mathcal{D}(\Omega)$ with $\operatorname{supp}\varphi \subset K$, there holds

$$-\|\varphi\|_{L^\infty(\Omega)}\Phi_K(x) \leq \varphi(x) \leq \|\varphi\|_{L^\infty(\Omega)}\Phi_K(x), \text{ in } \Omega,$$

i.e.,

$$\varphi(x) + \|\varphi\|_{L^\infty(\Omega)}\Phi_K(x), \ \|\varphi\|_{L^\infty(\Omega)}\Phi_K(x) - \varphi(x) \geq 0, \text{ in } \Omega.$$

From assumption (2.3.21) or (2.3.22), there hold

$$< f^\varepsilon, \varphi + \|\varphi\|_{L^\infty(\Omega)}\Phi_K > \leq 0 \tag{2.3.26}$$

and

$$< f^\varepsilon, \|\varphi\|_{L^\infty(\Omega)}\Phi_K - \varphi > \leq 0. \tag{2.3.27}$$

Since $\{f^\varepsilon\}_{\varepsilon>0}$ is weakly convergent in $W^{-1,p}(\Omega)$, hence $\{f^\varepsilon\}_{\varepsilon>0}$ is bounded in $W^{-1,p}(\Omega)$, (2.3.26) and (2.3.27) imply

$$|<f^\varepsilon,\varphi>|\leq <f^\varepsilon,\Phi_K>\|\varphi\|_{L^\infty(\Omega)}\leq C_K\|\varphi\|_{L^\infty(\Omega)}.$$

Step 2: Suppose $\Phi\in\mathcal{D}(\Omega)$ and $\Omega'\subset\Omega$ is a bounded domain, such that $\operatorname{supp}\Phi\subset\Omega'$. If Ω is bounded, choose $\Omega'=\Omega$. From assumptions (2.3.20) and (2.3.21), (2.3.25) implies

$$\Phi f^\varepsilon\rightharpoonup\Phi f^0 \text{ weakly in } W^{-1,p}(\Omega), \quad\text{as } \varepsilon\to 0; \tag{2.3.28}$$

and for any $\varphi\in\mathcal{D}(\Omega)$, $\varphi\geq 0$, there holds

$$|<\Phi f^\varepsilon,\varphi>|\leq M\|\varphi\|_{L^\infty(\Omega)}, \tag{2.3.29}$$

where M is a positive constant independent of φ.

Let $C_0(\bar\Omega')$ denote the space of functions which are continuous on $\bar\Omega'$ and vanish on $\partial\Omega'$, with norm $\|\cdot\|_{L^\infty(\bar\Omega')}$. Then $C_0(\bar\Omega')$ is a Banach space, and (2.3.29) implies that

$$\Phi f^\varepsilon \text{ is bounded in } (C_0(\bar\Omega'))', \tag{2.3.30}$$

where $(C_0(\bar\Omega'))'$ denotes the dual space of $C_0(\bar\Omega')$.

Step 3: From the Sobolev imbedding theorem, for any $r>N$, imbedding $W_0^{1,r}(\Omega')\hookrightarrow C_0(\bar\Omega')$ is compact. Then by the duality principle, imbedding $(C_0(\bar\Omega'))'\hookrightarrow W^{-1,r'}(\Omega')$ is compact for $1\leq r'=\frac{r}{(r-1)}\leq\frac{N}{N-1}$. Thus (2.3.30) implies that: for any $1\leq r'\leq\frac{N}{N-1}$, there holds

$$\Phi f^\varepsilon\to\Phi f^0 \text{ strongly in } W^{-1,r'}(\Omega'), \quad\text{as } \varepsilon\to 0. \tag{2.3.31}$$

Step 4: From the classical interpolation theorem, for $0<\theta<1$, $1<p_0$, $p_1<+\infty$, $\frac{1}{q}=\frac{1-\theta}{p_0}+\frac{\theta}{p_1}$, there holds

$$\left(W^{-1,p_0}(R^N),W^{-1,p_1}(R^N)\right)_{\theta,q}=W^{-1,q}(R^N).$$

Particularly, if $\varphi\in W^{-1,p_0}(R^N)\cap W^{-1,p_1}(R^N)$, then $\varphi\in W^{-1,q}(R^N)$ and

$$\|\varphi\|_{W^{-1,q}(R^N)}\leq C\|\varphi\|_{W^{-1,p_0}(R^N)}^{1-\theta}\|\varphi\|_{W^{-1,p_1}(R^N)}^\theta. \tag{2.3.32}$$

Let $p_0 = p$, $p_1 = r'$ for some $1 < r < N/(N-1)$ and extend $\Phi f^\varepsilon = 0$ outside Ω'. Then for all $1 < q < p$, (2.3.32) implies

$$\|\Phi f^\varepsilon - \Phi f^0\|_{W^{-1,q}(R^N)}$$

$$\leq C\|\Phi f^\varepsilon - \Phi f^0\|_{W^{-1,p_0}(R^N)}^{1-\theta} \|\Phi f^\varepsilon - \Phi f^0\|_{W^{-1,p_1}(R^N)}^{\theta}.$$

Then (2.3.31) and (2.3.28) imply (2.3.23).

Step 5: For any $\eta > 0$ small enough, $\psi^\eta \in \mathcal{D}(\Omega)$ is the function obtained in Lemma 2.3.3, i.e., ψ^η satisfies (2.3.10) and $(1 - \psi^\eta)f^\varepsilon \in W^{-1,p}(\Omega)$. Then for any $\varphi \in \mathcal{D}(\Omega)$, there holds

$$|< (1 - \psi^\eta)f^\varepsilon, \varphi >| \quad = |< f^\varepsilon, (1 - \psi^\eta)\varphi >|$$

$$\leq \|f^\varepsilon\|_{W^{-1,p}(\Omega)} \|(1 - \psi^\eta)\varphi\|_{W_0^{1,p'}(\Omega)}$$

$$\leq \eta \|f^\varepsilon\|_{W^{-1,p}(\Omega)} \|\varphi\|_{W_0^{1,q'}(\Omega)}, \quad q' > p',$$

i.e.,

$$\|(1 - \psi^\eta)f^\varepsilon\|_{W^{-1,q}(\Omega)} \leq \eta \|f^\varepsilon\|_{W^{-1,p}(\Omega)}.$$

For $\eta > 0$ fixed, from (2.3.23), there holds

$$\psi^\eta f^\varepsilon \to \psi^\eta f^0 \text{ strongly in } W^{-1,q}(\Omega'), \quad \text{as } \varepsilon \to 0, \ \forall \, q < p.$$

On the other hand, there exists decomposition

$$f^\varepsilon = (1 - \psi^\eta)f^\varepsilon + \psi^\eta f^\varepsilon, \ f^0 = (1 - \psi^\eta)f^0 + \psi^\eta f^0.$$

Let $\eta \to 0$. Then there holds

$$f^\varepsilon \to f^0 \text{ strongly in } W^{-1,q}(\Omega), \quad \text{as } \varepsilon \to 0, \ \forall \, q < p.$$

This completes the proof of Theorem 2.3.4. ∎

Chapter 3

Cauchy Problem for Scalar Equation

In this chapter, we shall consider the Cauchy problem for the scalar conservation law ($n = 1$). The study of the single equation has a long history and the existence and uniqueness of the generalized solution for this equation has been well studied (cf. [Ho, Kr, La1, Ol] or the references cited in [Sm]). As the simplest model of hyperbolic conservation laws, Tartar first introduced the theory of compensated compactness to the scalar equation and succeeded in obtaining a new method, called the compensated compactness method, to study the global existence of generalized solutions for hyperbolic conservation laws. The original proof of Tartar is given in [Ta]. In this chapter, we use a simplified proof from [CL1, Lu1] to study both L^∞ (Sec. 3.1) and $L^p, 1 < p < \infty$ (Sec. 3.2) solutions, in which two pairs of entropy-entropy flux ((3.1.4), (3.1.5)) and the weak continuity theorem of a 2×2 determinant (Theorem 2.1.4) play a more important role, but the idea of Young measures (Theorem 2.2.1) has been avoided.

3.1 L^∞ Solution

In this section, we shall give a simple proof of existence of global generalized solutions to the Cauchy problem of scalar conservation law:

$$\begin{cases} u_t + f(u)_x = 0, \\ u(x, 0) = u_0(x), \end{cases} \tag{3.1.1}$$

where the real-valued function f is of class C^2 and the initial data $u_0(x)$ is bounded measurable.

Theorem 3.1.1 *Let $\Omega \subset R \times R^+$ be a bounded open set and $u^\varepsilon : \Omega \to R$ be a sequence of functions such that*

$$w^\star - \lim u^\varepsilon = u, \qquad w^\star - \lim f(u^\varepsilon) = v, \qquad (3.1.2)$$

where $f \in C^2(-\|u_0(x)\|_{L^\infty}, \|u_0(x)\|_{L^\infty})$. Suppose that

$$\eta_i(u^\varepsilon)_t + q_i(u^\varepsilon)_x \qquad \text{lies in a compact set of } H^{-1}_{loc}(\Omega), (i = 1, 2) \qquad (3.1.3)$$

where

$$(\eta_1(\theta), q_1(\theta)) = (\theta - k, f(\theta) - f(k)) \qquad (3.1.4)$$

and

$$(\eta_2(\theta), q_2(\theta)) = \left(f(\theta) - f(k), \int_k^\theta (f'(s))^2 ds\right), \qquad (3.1.5)$$

where k is an arbitrary constant. Then,
    (1) $v = f(u)$, a.e., for any $f \in C^2$ and
    (2) $u^\varepsilon \to u$, a.e., if

$$meas \ \{u : f''(u) = 0\} = 0. \qquad (3.1.6)$$

Proof.   Using (3.1.3) and the weak continuity theorem of a 2×2 determinant (Theorem 2.1.4), we have

$$w^\star - \lim \begin{vmatrix} \eta_1(u^\varepsilon) & q_1(u^\varepsilon) \\ \eta_2(u^\varepsilon) & q_2(u^\varepsilon) \end{vmatrix} = \begin{vmatrix} \overline{\eta_1(u^\varepsilon)} & \overline{q_1(u^\varepsilon)} \\ \overline{\eta_2(u^\varepsilon)} & \overline{q_2(u^\varepsilon)} \end{vmatrix} \qquad (3.1.7)$$

here, and hereafter the weak-star limit is denoted by $w^\star - \lim \eta(u^\varepsilon) = \overline{\eta(u^\varepsilon)}$ and $w^\star - \lim q(u^\varepsilon) = \overline{q(u^\varepsilon)}$.
    But since

$$\begin{vmatrix} \overline{\eta_1(u^\varepsilon)} & \overline{q_1(u^\varepsilon)} \\ \overline{\eta_2(u^\varepsilon)} & \overline{q_2(u^\varepsilon)} \end{vmatrix} = \overline{u^\varepsilon - k} \int_k^{\overline{u^\varepsilon}} (f'(s))^2 ds - (\overline{f(u^\varepsilon)} - f(k))^2, \qquad (3.1.8)$$

and

$$
\begin{vmatrix} \eta_1(u^\varepsilon) & q_1(u^\varepsilon) \\ \eta_2(u^\varepsilon) & q_2(u^\varepsilon) \end{vmatrix}
$$

$$
= (u^\varepsilon - u) \int_u^{u^\varepsilon} (f'(s))^2 ds - (f(u^\varepsilon) - f(u))^2 \tag{3.1.9}
$$

$$
+ (u - k) \int_u^{u^\varepsilon} (f'(s))^2 ds + (u^\varepsilon - k) \int_k^u (f'(s))^2 ds
$$

$$
- (f(u) - f(k))^2 - 2(f(u^\varepsilon) - f(u))(f(u) - f(k)),
$$

then there exists a zero measure set Ω_1, for any $(x, t) \in \Omega - \Omega_1$, from (3.1.7)-(3.1.9), we have

$$
\overline{(u^\varepsilon - u) \int_u^{u^\varepsilon} (f'(s))^2 ds - (f(u^\varepsilon) - f(u))^2}
$$

$$
+ (u - k) \overline{\int_u^{u^\varepsilon} (f'(s))^2 ds} + \overline{(u^\varepsilon - k)} \int_k^u (f'(s))^2 ds
$$

$$
- (f(u) - f(k))^2 - 2\overline{(f(u^\varepsilon) - f(u))}(f(u) - f(k)) \tag{3.1.10}
$$

$$
= \overline{u^\varepsilon - k} \overline{\int_k^{u^\varepsilon} (f'(s))^2 ds} - \overline{(f(u^\varepsilon) - f(k))^2}
$$

$$
= \overline{u^\varepsilon - k} \overline{\int_k^{u^\varepsilon} (f'(s))^2 ds} - \overline{(f(u^\varepsilon) - f(u))^2}
$$

$$
- (f(u) - f(k))^2 - 2\overline{(f(u^\varepsilon) - f(u))}(f(u) - f(k)).
$$

Since

$$\overline{u^\varepsilon - k \int_k^{u^\varepsilon} (f'(s))^2 ds}$$

$$= \overline{(u^\varepsilon - u) \int_u^{u^\varepsilon} (f'(s))^2 ds}$$

$$+ \overline{(u - k) \int_u^{u^\varepsilon} (f'(s))^2 ds} + \overline{u^\varepsilon - k} \int_k^{u} (f'(s))^2 ds$$

$$= \overline{(u - k) \int_u^{u^\varepsilon} (f'(s))^2 ds} + \overline{u^\varepsilon - k} \int_k^{u} (f'(s))^2 ds,$$

$(3.1.11)$

we have from $(3.1.10)$ and $(3.1.11)$ that

$$\overline{(u^\varepsilon - u) \int_u^{u^\varepsilon} (f'(s))^2 ds - (f(u^\varepsilon) - f(u))^2} + (\overline{f(u^\varepsilon) - f(u)})^2 = 0.$$

$(3.1.12)$

Since both terms in the left-hand side of $(3.1.12)$ are nonnegative, we have

$$\overline{(u^\varepsilon - u) \int_u^{u^\varepsilon} (f'(s))^2 ds - (f(u^\varepsilon) - f(u))^2} = 0 \qquad (3.1.13)$$

and

$$(\overline{f(u^\varepsilon) - f(u)})^2 = 0. \qquad (3.1.14)$$

So $v = f(u)$ is obtained by $(3.1.14)$. It follows from $(3.1.13)$ that

$$\lim_{\varepsilon \to 0} \int_\Omega (u^\varepsilon - u) \int_u^{u^\varepsilon} (f'(s))^2 ds - (f(u^\varepsilon) - f(u))^2 dx dt = 0 \qquad (3.1.15)$$

and hence,

$$\lim_{\varepsilon \to 0} \int_{\Omega(|u^\varepsilon - u| > \alpha)} (u^\varepsilon - u) \int_u^{u^\varepsilon} (f'(s))^2 ds - (f(u^\varepsilon) - f(u))^2 dx dt = 0.$$

$(3.1.16)$

Since

$$\frac{d}{d\theta}\left((\theta - u)\int_u^\theta (f'(s))^2 ds - (f(\theta) - f(u))^2\right) = \int_u^\theta (f'(\theta) - f'(s))^2 ds,$$

(3.1.17)

and if $f''(u) \neq 0, a.e.,$ then

$$\int_{\Omega((u^\varepsilon - u) > \alpha)} (u^\varepsilon - u)\int_u^{u^\varepsilon} (f'(s))^2 ds - (f(u^\varepsilon) - f(u))^2 dx dt$$

$$\geq C_\alpha \mathrm{meas}\left(\Omega((u^\varepsilon - u) > \alpha)\right)$$

(3.1.18)

and

$$\int_{\Omega((u^\varepsilon - u) < -\alpha)} (u^\varepsilon - u)\int_u^{u^\varepsilon} (f'(s))^2 ds - (f(u^\varepsilon) - f(u))^2 dx dt$$

$$\geq C_\alpha \mathrm{meas}\left(\Omega((u^\varepsilon - u) < -\alpha)\right)$$

(3.1.19)

for a suitable positive constant C_α, which is independent of ε. Therefore for any given constant α, we have

$$\lim_{\varepsilon \to 0} \mathrm{meas}\left(\Omega(|u^\varepsilon - u| > \alpha)\right) = 0,$$

(3.1.20)

which implies the pointwise convergence of a subsequence of u^ε on Ω. ∎

Theorem 3.1.2 *If $u_0(x) \in L^\infty, f \in C^2(-\|u_0(x)\|_{L^\infty}, \|u_0(x)\|_{L^\infty}),$ then the sequence of viscosity solutions $\{u^\varepsilon\}$ uniquely defined by the Cauchy problem*

$$(u^\varepsilon)_t + f(u^\varepsilon)_x = \varepsilon(u^\varepsilon)_{xx}$$

(3.1.21)

with bounded measurable initial data

$$u(x, 0) = u_0(x)$$

(3.1.22)

satisfies the compactness (3.1.3).

Proof. Since the initial data (3.1.22) is bounded in L^∞ space, by the standard maximum principle of parabolic equations, the viscosity solutions u^ε have an *a priori* L^∞ estimate

$$||u^\varepsilon(x,t)||_{L^\infty} \leq ||u_0(x)||_{L^\infty}, \qquad (3.1.23)$$

which implies the existence of u^ε for $t > 0$ (see Theorem 1.0.2).

Let $K \subset R \times R^+$ be an arbitrary compact set and choose $\phi \in C_0^\infty(R \times R^+)$ such that $\phi_K = 1, 0 \leq \phi \leq 1$.

Multiplying Equation (3.1.21) by $u^\varepsilon \phi$ and integrating over $R \times R^+$, we obtain

$$\varepsilon \int_0^\infty \int_{-\infty}^\infty (u_x^\varepsilon)^2 \phi \, dx \, dt$$

$$= \int_0^\infty \int_{-\infty}^\infty \left[\frac{1}{2}(u^\varepsilon)^2 \phi_t + \left(u^\varepsilon f(u^\varepsilon) - \int_0^{u^\varepsilon} f(s) ds\right) \phi_x \right. \qquad (3.1.24)$$

$$\left. + \frac{1}{2}\varepsilon(u^\varepsilon)^2 \phi_{xx} \right] dx \, dt \leq M(\phi),$$

and hence that

$$\varepsilon(u_x^\varepsilon)^2 \text{ is bounded in } L^1_{loc}(R \times R^+). \qquad (3.1.25)$$

For any entropy $\eta \in C^2$, we have from Equation (3.1.21) that

$$\eta(u^\varepsilon)_t + q(u^\varepsilon)_x = \varepsilon\eta(u^\varepsilon)_{xx} - \varepsilon\eta''(u^\varepsilon)(u_x^\varepsilon)^2 = I_1 + I_2, \qquad (3.1.26)$$

where q is the entropy flux corresponding to the entropy η. by (3.1.25), it is easy to see that I_2 is bounded in $L^1_{loc}(R \times R^+)$, and hence compact in $W^{-1,\alpha}_{loc}$, for some exponent $\alpha \in (1,2)$; and I_1 is compact in H^{-1}_{loc}. The left-hand side of (3.1.26) is bounded in $W^{-1,\infty}_{loc}$ because of the boundedness of u^ε in L^∞. So Theorem 3.1.2 is proved by the embedding Theorem 2.3.2. ∎

Combining Theorem 3.1.1 and Theorem 3.1.2, we get the following main theorem in this section:

Theorem 3.1.3 *If $u_0(x) \in L^\infty, f \in C^2(-||u_0(x)||_{L^\infty}, ||u_0(x)||_{L^\infty})$, then the Cauchy problem (3.1.1) has a weak solution $u \in L^\infty$ in the* sense of Definition 1.0.1. Moreover, if f satisfies the condition (3.1.6), then the weak solution u is also the entropy solution in the sense of Lax, namely

$$\int_0^\infty \int_{-\infty}^\infty \eta(u)\phi_t + q(u)\phi_x \, dx \, dt \geq 0, \tag{3.1.27}$$

where the C^2 function pair (η, q) satisfies $q' = \eta' f', \eta'' \geq 0$ and $\phi \in C_0^\infty(R \times R^+ - \{t = 0\})$ is a positive function.

3.2 L^p Solution, $1 < p < \infty$

In this section, we consider the Cauchy problem (3.1.1) again, but the nonlinear flux function and the initial data satisfy the following conditions:

(C_1) $f(u) \in C^2$, and satisfies

$$|f(u)| \leq C|u|^{s+1}, \quad |f'(u)| \leq C|u|^s, \quad (s \geq 0). \tag{3.2.1}$$

(C_2) $\|u_0(x)\|_{2(s+1)} \lesssim M$.

Similar to Section 3.1, we introduce the following parabolic equation:

$$u_t^\varepsilon + f(u^\varepsilon)_x = \varepsilon u_{xx}^\varepsilon \tag{3.2.2}$$

with the initial data

$$u^\varepsilon(x, 0) = u_0(x) * G^\varepsilon = u_0^\varepsilon(x), \tag{3.2.3}$$

where G^ε is a mollifier such that $u_0^\varepsilon(x)$ are smooth,

$$\lim_{|x| \to \infty} u_0^\varepsilon(x) = 0, \quad u_0^\varepsilon(x) \to u_0(x) \text{ a.e., as } \varepsilon \to 0, \tag{3.2.4}$$

and

$$\|u_0^\varepsilon(x)\|_{L^\infty} \leq M(\varepsilon), \quad \|u_0^\varepsilon(x)\|_{2(s+1)} \leq \|u_0(x)\|_{2(s+1)} \leq M, \tag{3.2.5}$$

for a positive constant $M(\varepsilon)$ depending on ε.

Theorem 3.2.1 *Let C_1, C_2 hold. Then for any fixed ε and the time T, there is a smooth solution u^ε of the Cauchy problem (3.2.2), (3.2.3) on $R \times (0, T]$ such that $u^\varepsilon \in C^\infty(R \times (0, T])$,*

$$||u(\cdot, t)||_{2(s+1)} \leq M, \quad \lim_{|x| \to \infty} u^\varepsilon(x, t) = 0 \qquad (3.2.6)$$

uniformly for $t \in [0, T]$.

Proof. Since the initial data $|u_0^\varepsilon(x)| \leq M(\varepsilon)$, then the existence is obtained similarly to Theorem 3.1.2. The solutions $u^\varepsilon(x, t) \to 0$ as $|x| \to \infty$ can be also obtained by Theorem 1.0.2. To prove the uniform $L^{2(s+1)}$ bound in the first part of (3.2.6), multiplying Equation (3.2.2) by $2(s+1)(u^\varepsilon)^{2s+1}$ and then integrating in $R \times R^+$, we get

$$\int_{-\infty}^{\infty} (u^\varepsilon)^{2(s+1)} dx + \int_0^\infty \int_{-\infty}^{\infty} \varepsilon 2(s+1)(2s+1)(u^\varepsilon)^{2s}(u_x^\varepsilon)^2 dx dt$$

$$\leq \int_{-\infty}^{\infty} (u_0^\varepsilon(x))^{2(s+1)} dx \leq M,$$

$$(3.2.7)$$

which completes the proof of Theorem 3.2.1. ∎

Now we study the $L^{2(s+1)}$ weak solution or generalized solution for the Cauchy problem (3.1.1).

Lemma 3.2.2 *Let C_1, C_2 hold. Then the solutions of the Cauchy problem (3.2.2), (3.2.3) satisfy:*

$$\varepsilon^{\frac{1}{2}} \partial_x u^\varepsilon \text{ are uniformly bounded in } L_{loc}^2(R \times (0, \infty)). \qquad (3.2.8)$$

Proof. Similar to the proof of (3.1.25) in Theorem 3.1.2, let $K \subset S \subset R \times (0, \infty)$ and choose $\phi \in C_0^\infty(R \times R^+)$ such that $\phi_K = 1, 0 \leq \phi \leq 1$ and $S = \text{supp}\,\phi$.

Multiplying Equation (3.2.2) by $u^\varepsilon \phi$ and integrating over $R \times R^+$, we obtain

$$\varepsilon \int_0^\infty \int_{-\infty}^{\infty} (u_x^\varepsilon)^2 \phi dx dt$$

$$= \int_0^\infty \int_{-\infty}^{\infty} \left[\frac{1}{2}(u^\varepsilon)^2 \phi_t + (u^\varepsilon f(u^\varepsilon) - \int_0^{u^\varepsilon} f(s) ds) \phi_x \right. \qquad (3.2.9)$$

$$\left. + \frac{1}{2}\varepsilon (u^\varepsilon)^2 \phi_{xx} \right] dx dt \leq M(\phi, ||u^\varepsilon||_{2(s+1)}),$$

which implies the proof of Lemma 3.2.2. ∎

Lemma 3.2.3 *Let C_1, C_2 hold. Then for any fixed n,*

$$\frac{\partial}{\partial t} I_n(u^\varepsilon) + \frac{\partial}{\partial x} f_n(u^\varepsilon), \quad \frac{\partial}{\partial t} f_n(u^\varepsilon) + \frac{\partial}{\partial x} F_n(u^\varepsilon) \tag{3.2.10}$$

lie in a compact set of $H_{loc}^{-1}(\Omega)$, where $\Omega \subset R \times R^+$ is any open and bounded set, and functions I_n, f_n, F_n are defined as follows:

$$I_n(u) = u, \ \text{if} \ |u| \leq n, \quad I_n = 0, \ \text{if} \ |u| \geq 2n,$$

$$I_n \in C^2, \quad |I_n(u)| \leq |u|, \quad |I_n'(u)| \leq 2,$$

$$f_n(u) = \int_0^u I_n'(s) f'(s) ds$$

and

$$F_n(u) = \int_0^u f_n'(s) f'(s) ds.$$

Proof. Using Lemma 3.2.2 and noticing the boundedness of these entropy pairs as n fixed, we can get the proof of Lemma 3.2.3 in a fashion similar to Theorem 3.1.2. ∎

Lemma 3.2.4 *Let C_1, C_2 hold. If f satisfies the condition (3.1.6), then there exists a subsequence (still labelled) u^ε such that*

$$u^\varepsilon \to u \ \text{strongly in} \ L^h \ \text{for all} \ h, \ 0 < h < 2(s+1). \tag{3.2.11}$$

Proof. From the boundedness of u^ε in $L^{2(s+1)}$, there exists a subsequence (still labelled) u^ε such that

$$u^\varepsilon \rightharpoonup u, \ \text{weakly in} \ L^{2(s+1)}. \tag{3.2.12}$$

Let $\overline{v^\varepsilon}$ express $v^\varepsilon \rightharpoonup \overline{v^\varepsilon}$ in the sense of distributions. Then using the weak continuity theorem of a 2×2 determinant (Theorem 2.1.4), we have

$$\overline{H_n(u^\varepsilon, k)} = H_n^*(u^\varepsilon, k), \tag{3.2.13}$$

where k is an arbitrary constant and

$$H_n(u^\varepsilon, k) = (F_n(u^\varepsilon) - F_n(k))(I_n(u^\varepsilon) - I_n(k)) - (f_n(u^\varepsilon) - f_n(k))^2, \tag{3.2.14}$$

$$H_n^\star(u^\varepsilon, k) = (\overline{F_n(u^\varepsilon)} - F_n(k))(\overline{I_n(u^\varepsilon)} - I_n(k)) - (\overline{f_n(u^\varepsilon)} - f_n(k))^2. \tag{3.2.15}$$

So, there exists a set Ω_1 of measure zero, such that

$$\overline{H_n(u^\varepsilon(x,t), u(y,\tau))} = H_n^\star(u^\varepsilon(x,t), u(y,\tau)) \text{ for any } (y,\tau) \in \Omega_1^c, \tag{3.2.16}$$

where Ω_1^c is the complement of Ω_1 in Ω. Thus

$$\overline{H_n(u^\varepsilon(x,t), u(x,t))} H_n^\star(u^\varepsilon(x,t), u(x,t)), \tag{3.2.17}$$

if choosing $(y, \tau) = (x, t) \in \Omega_1^c$.

Since

$$\int_\Omega |H_n(u^\varepsilon(x,t), u(x,t))| dx$$
$$\tag{3.2.18}$$
$$\leq C + C \int_\Omega |u^\varepsilon(x,t)|^{2(s+1)} + |u(x,t)|^{2(s+1)} dx \leq C_1$$

and

$$\int_\Omega |H_n^\star(u^\varepsilon(x,t), u(x,t))| dx$$
$$\tag{3.2.19}$$
$$\leq C + C \int_\Omega |u^\varepsilon(x,t)|^{2(s+1)} + |u(x,t)|^{2(s+1)} dx \leq C_1,$$

we have by Lebesgue's dominant convergence theorem that

$$\lim_{n \to \infty} H_n^\star(u^\varepsilon, u) = -(\overline{f(u^\varepsilon)} - f(u))^2. \tag{3.2.20}$$

So we have from (3.2.14) that

$$(u^\varepsilon - u) \int_u^{u^\varepsilon} (f'(s))^2 ds - (f(u^\varepsilon) - f(u))^2 = -(\overline{f(u^\varepsilon)} - f(u))^2. \tag{3.2.21}$$

To satisfy

$$(u^\varepsilon - u) \int_u^{u^\varepsilon} (f'(s))^2 ds - (f(u^\varepsilon) - f(u))^2 \geq 0,$$

we have

$$\overline{f(u^\varepsilon)} = f(u) \tag{3.2.22}$$

and

$$\overline{(u^\varepsilon - u) \int_u^{u^\varepsilon} (f'(s))^2 ds - (f(u^\varepsilon) - f(u))^2} = 0. \tag{3.2.23}$$

The left is exactly analogous to that of Theorem 3.1.1. Thus we get the proof of Lemma 3.2.4. ∎

Combining Lemmas 3.2.1-3.2.4, we have the following main theorem in this section:

Theorem 3.2.5 *Let C_1, C_2 hold. If f satisfies meas $\{u : f''(u) = 0\} = 0$, then the Cauchy problem (3.1.1) has an $L^{2(s+1)}$ weak solution defined by Definition 1.0.1.*

3.3 Related Results

After Tartar's proof for scalar equation in L^∞ space, Schonbek [SC] extended the compensated compactness method to study the strong convergence of the sequence of $L^p, 1 < p < \infty$ solutions $u^{\varepsilon,\delta}$ for the Cauchy problem of the following Korteweg de Vries equation with viscosity

$$u_t + uu_x + \delta u_{xxx} = \varepsilon u_{xx}, \tag{3.3.1}$$

and the generalized BBM-Burger equation with viscosity

$$u_t + uu_x - \delta u_{xxt} = \varepsilon u_{xx}. \tag{3.3.2}$$

Under suitable order relations between ε and δ, and the strictly convex condition on the nonlinear flux function f, Schonbek succeeded in obtaining an L^p weak solution for the Cauchy problem (3.1.1).

The restriction of the strict convexity on the flux function f in Schonbek's paper is removed by Lu in [Lu1]. In Lu's proof, the ideas of Young measures are also avoided.

In [Lu3, Lu7], this simplified proof is used to study weak solutions to the special system of two equations,

$$\begin{cases} (u + qz)_t + f(u)_x = 0, \\ z_t + kg(u)z = 0, \end{cases} \tag{3.3.3}$$

which is derived by Majda [Ma] as a chemical reaction model, where q is a positive constant, k denotes the reaction rate of the chemical material.

In [Sz], Szepessy extends the concept of measure valued solution, initially introduced by DiPerna [Di4], to the scalar conservation law in the case of several space variables. He obtains the unique $L^p, p > 1$ weak solution without the condition (3.1.6), i.e., meas $\{u : f''(u) = 0\} = 0$ on the flux function f. In his proof, the growth condition (3.2.1) on f is extended to $|f(u)| \le C|u|^q$ for any $0 < q < p$ if the initial data is bound in L^p space.

Chapter 4

Preliminaries in 2×2 Hyperbolic System

Besides the scalar equation studied in Chapter 3, a major part in this book focuses on the applications on hyperbolic conservation systems of two equations. In the next several chapters (Chapters 5-12), we shall apply the compensated compactness method to different types of systems of two equations. Here the types are distinguished by hyperbolicity, linearity and so on.

4.1 Basic Definitions

We consider the pair of conservation laws

$$u_t + f(u,v)_x = 0, \quad v_t + g(u,v)_x = 0, \tag{4.1.1}$$

where u and v are in R. We let $U = (u,v)$ and $F(U) = (f,g)$ so that the equations in (4.1.1) can be written as

$$U_t + dF(U)U_x = 0, \tag{4.1.2}$$

where $dF(U)$ is the Jacobian matrix of F. The following definitions can be found from Smoller's book [Sm].

Definition 4.1.1 *We say that system (4.1.2) is hyperbolic if dF has two real eigenvalues λ_1 and λ_2. System (4.1.2) is called strictly hyperbolic if λ_1 and λ_2 are distinct, i.e., $\lambda_1 < \lambda_2$. If λ_1 and λ_2 coincide at* some points or domains, system (4.1.2) is called nonstrictly hyperbolic or hyperbolically degenerate.

Let $l_{\lambda_1}, l_{\lambda_2}, r_{\lambda_1}$ and r_{λ_2} denote the corresponding left and right eigenvectors.

Definition 4.1.2 *We say that (4.1.2) is genuinely nonlinear in the λ_1 characteristic field if*

$$\nabla \lambda_1 \cdot r_{\lambda_1} \neq 0 \tag{4.1.3}$$

and genuinely nonlinear in the λ_2 characteristic field if

$$\nabla \lambda_2 \cdot r_{\lambda_2} \neq 0. \tag{4.1.4}$$

If $\nabla \lambda_1 \cdot r_{\lambda_1} = 0$ or $\nabla \lambda_2 \cdot r_{\lambda_2} = 0$ at some domain D, then system (4.1.2) is called linearly degenerate in D in the λ_1 characteristic field or in the λ_2 characteristic field.

Definition 4.1.3 *The functions $w = w(u,v), z = z(u,v)$ are called Riemann invariants of system (4.1.2) corresponding to λ_1 and λ_2 if they satisfy the equations*

$$\nabla w \cdot r_{\lambda_1} = 0, \quad \nabla z \cdot r_{\lambda_2} = 0. \tag{4.1.5}$$

Definition 4.1.4 *A pair of functions $(\eta(u), q(u))$ is called a pair of entropy-entropy flux of system (4.1.2) if $(\eta(u), q(u))$ satisfies*

$$\nabla q(u) = \nabla \eta(u) \nabla f(u). \tag{4.1.6}$$

If $n \leq 2$, we can always resolve system (4.1.6) and obtain a class of entropies and the corresponding entropy fluxes. However, if $n > 2$, system (4.1.6) is over-determined and could be resolved only for some very special cases. As we have seen in Chapter 3, the applications of the compensated compactness method on conservation laws depend strongly on the constructions of entropies of hyperbolic systems we considered. This is why, until now, almost all the results obtained by this method are only for systems of two equations.

4.2 L^∞ Estimate of Viscosity Solutions

Similar to the scalar equation, for a given hyperbolic system, we first need to construct its approximation solutions, for instance, the sequence of viscosity solutions given by the Cauchy problem

$$\begin{cases} u_t + f(u,v)_x = \varepsilon u_{xx}, \\ v_t + g(u,v)_x = \varepsilon v_{xx}, \end{cases} \tag{4.2.1}$$

with bounded measurable initial data

$$(u(x,0), v(x,0)) = (u_0(x), v_0(x)). \tag{4.2.2}$$

To obtain an *a priori* uniform L^∞ estimate of $(u^\varepsilon, v^\varepsilon)$ with respect to the viscosity parameter ε, in general, the unique framework we could use is found by Chueh, Conley and Smoller [CCS], and is called the Invariant Region Method. We summarize the main result of [CCS] about the solutions of the Cauchy problem (4.2.1), (4.2.2) in the following theorem:

Theorem 4.2.1 *Let w, z be two Riemann invariants of system (4.1.1). If the curve $w = M_1$ (or $z = M_2$) for a constant M_1 (or M_2) in the (u,v)-plane is upper-convex, i.e., $w_{uu}a^2 + 2w_{uv}ab + w_{vv}b^2 \geq 0$ or $z_{uu}a^2 + 2z_{uv}ab + z_{vv}b^2 \geq 0$ for a vector $(a,b) \in R^2$, then the solutions $(u^\varepsilon(x,t), v^\varepsilon(x,t))$ satisfy*

$$w(u^\varepsilon, v^\varepsilon) \leq M_1 \tag{4.2.3}$$

or

$$z(u^\varepsilon, v^\varepsilon) \leq M_2 \tag{4.2.4}$$

if the initial data $(u_0(x), v_0(x))$ satisfy the same estimate $w(u_0, v_0) \leq M_1$ or $z(u_0, v_0) \leq M_2$.

If the curves $w = M_1$ and $z = M_2$ both are upper-convex, then both estimates (4.2.3) and (4.2.4) are true if the initial data satisfy the same estimates.

Chapter 5

A Symmetry System

In this chapter, we are concerned with a hyperbolic system of two equations with symmetry

$$\begin{cases} u_t + (u\phi(r))_x = 0, \\ v_t + (v\phi(r))_x = 0 \end{cases} \tag{5.0.1}$$

with bounded measurable initial data

$$(u(x,0), v(x,0)) = (u_0(x), v_0(x)), \tag{5.0.2}$$

where $\phi(r)$ is a nonlinear symmetric function of u, v with $r = u^2 + v^2$. System (5.0.1) is of interest because it arises in such areas as elasticity theory, magnetohydrodynamics, and enhanced oil recovery (cf. [KK, LW]).

Let F be the mapping from R^2 into R^2 defined by

$$F : (u, v) \rightarrow (u\phi(r), v\phi(r)).$$

Then two eigenvalues of dF are

$$\lambda_1 = \phi(r), \quad \lambda_2 = \phi + 2r\phi'(r) \tag{5.0.3}$$

with corresponding right eigenvectors

$$r_1 = (-v, u)^T, \quad r_2 = (u, v)^T. \tag{5.0.4}$$

By simple calculations,

$$\nabla\lambda_1 \cdot r_1 = 0, \quad \nabla\lambda_2 \cdot r_2 = 6r\phi'(r) + 4r^2\phi''(r). \tag{5.0.5}$$

Therefore, from (5.0.3) the strict hyperbolicity of system (5.0.1) fails on the points where $r\phi'(r) = 0$, and from (5.0.5) the first characteristic field is always linearly degenerate and the second characteristic field is either genuinely nonlinear or linearly degenerate, depending on the behavior of ϕ.

In this chapter, we assume that

$$\phi \in C^2(R^+), \quad \text{meas } \{r : 3\phi'(r) + 2r\phi''(r) = 0\} = 0, \qquad (5.0.6)$$

which is similar to the condition in (3.1.6). Therefore the second characteristic field could be linearly degenerate on a set of Lebesgue measure zero.

Consider the Cauchy problem for the related parabolic system

$$\begin{cases} u_t + (u\phi(r))_x = \varepsilon u_{xx}, \\ v_t + (v\phi(r))_x = \varepsilon v_{xx} \end{cases} \qquad (5.0.7)$$

with the initial data (5.0.2).

We have the main result in this chapter:

Theorem 5.0.1 *(1) Let the initial data $(u_0(x), v_0(x))$ be bounded measurable. Then for fixed $\varepsilon > 0$, the viscosity solution $(u^\varepsilon(x,t), v^\varepsilon(x,t))$ of the Cauchy problem (5.0.7), (5.0.2) exists and is uniformly bounded with respect to the viscosity parameter ε.*

(2) Moreover, if condition (5.0.6) holds, then there exists a subsequence of $r^\varepsilon = (u^\varepsilon)^2 + (v^\varepsilon)^2$ (still labelled r^ε) which converges pointwisely to a function $l(x,t)$.

(3) If $v_0(x) \geq c_0 > 0$ for a constant c_0 and the total variation of $\dfrac{u_0(x)}{v_0(x)}$ is bounded in $(-\infty, \infty)$ or equivalently

$$\int_\infty^\infty |(\frac{u_0(x)}{v_0(x)})_x| dx \leq M, \qquad (5.0.8)$$

then there exists a subsequence of $(u^\varepsilon, v^\varepsilon)$ (still denoted by $(u^\varepsilon, v^\varepsilon)$) which converges pointwisely to a pair of functions $(u(x,t), v(x,t))$ satisfying $l(x,t) = u^2(x,t) + v^2(x,t)$, which combining with (2) implies that the limit (u, v) is a weak solution of the hyperbolic system (5.0.1) with the initial data (5.0.2).

Remark 5.0.2 *Since $(u^\varepsilon, v^\varepsilon)$ is uniformly bounded with respect to ε, its weak-star limit (u, v) always exists. However, the strong limit $l(x, t)$ of $(u^\varepsilon)^2 + (v^\varepsilon)^2$ needs not equal $u(x, t)^2 + v(x, t)^2$. If this equality is true, then, at least, (u, v) is a weak solution of (5.0.1), (5.0.2) without any more condition such as given in part (3).*

Remark 5.0.3 *Since one characteristic field of system (5.0.1) is always linearly degenerate, the further condition (5.0.8) in Theorem 5.0.1 about the derivative of the initial data seems necessary to ensure the strong convergence of the viscosity solutions $(u^\varepsilon, v^\varepsilon)$. This fact also can be seen by the following simple example:*

Consider the initial value problem for the scalar linear equation

$$\begin{cases} u_t = \varepsilon u_{xx}, \\ u(x, 0) = u_0^\varepsilon(x). \end{cases} \tag{5.0.9}$$

The solutions of (5.0.9) are

$$u^\varepsilon(x, t) = \int_{-\infty}^{\infty} u_0^\varepsilon(y) G(x - y, t) dy,$$

where $G(x-y, t)$ is the Green function. Then clearly $u^\varepsilon(x, t)$ is compact only with some extra compact conditions on the initial data.

The proof of Theorem 5.0.1 will be given in the next sections.

5.1 Viscosity Solutions

To prove the existence of the viscosity solution in Theorem 5.0.1, from Theorem 1.0.2, it is sufficient to get the uniform L^∞ bound.

Multiplying the first and second equations of the parabolic system (5.0.7) by $2u$ and $2v$, respectively, then adding the result, we have

$$r_t + r_x \phi(r) + 2r(\phi(r))_x = \varepsilon r_{xx} - 2\varepsilon(u_x^2 + v_x^2) \tag{5.1.1}$$

or

$$r_t + f(r)_x \leq \varepsilon r_{xx}, \tag{5.1.2}$$

where

$$f(r) = \int_0^r \phi(s) + 2s\phi'(s)ds. \tag{5.1.3}$$

Since the initial data is bounded, then $r(0, t)$ is bounded from above. Using the maximum principle to (5.1.2), we get $r = (u^\varepsilon)^2 + (v^\varepsilon)^2 \le M$, which implies the uniform boundedness of $(u^\varepsilon, v^\varepsilon)$ and hence, the existence of the viscosity solutions.

To prove the second part of Theorem 5.0.1, namely the strong convergence of r^ε, we multiply Equation (5.1.1) by a test function ϕ, where $\phi \in C_0^\infty(R \times R^+)$ satisfies $\phi_K = 1, 0 \le \phi \le 1$ and $S = \operatorname{supp} \phi$ for an arbitrary compact set $K \subset S \subset R \times R^+$. Then, similar to the proof of (3.1.25), we have that

$$\int_0^\infty \int_{-\infty}^\infty 2\varepsilon \big((u_x^\varepsilon)^2 + (v_x^\varepsilon)^2\big)\phi dxdt$$

$$= \int_0^\infty \int_{-\infty}^\infty \big(\varepsilon r_{xx} - r_t - f(r)_x\big)\phi dxdt \tag{5.1.4}$$

$$= \int_0^\infty \int_{-\infty}^\infty \varepsilon r\phi_{xx} + r\phi_t + f(r)\phi_x dxdt \le M(\phi)$$

and hence

$$\varepsilon(u_x^\varepsilon)^2 \text{ and } \varepsilon(v_x^\varepsilon)^2 \text{ are bounded in } L_{loc}^1(R \times R^+). \tag{5.1.5}$$

Let $(\eta(r), q(r))$ be any pair of entropy-entropy flux of the scalar equation

$$r_t + \left(\int_0^r \phi(s) + 2s\phi'(s)ds\right)_x = 0 \tag{5.1.6}$$

and multiply (5.1.1) by $\eta'(r)$. Then

$$\eta(r)_t + q(r)_x$$

$$= \varepsilon(\eta'(r)r_x)_x - \varepsilon\eta''(r)r_x^2 - 2\varepsilon\eta'(r)(u_x^2 + v_x^2) \tag{5.1.7}$$

$$= I_1 - I_2 - I_3,$$

where I_1 is compact in $W_{loc}^{-1,2}$ and $I_2 + I_3$ are bounded in $L_{loc}^1(R \times R^+)$, and hence compact in $W_{loc}^{-1,\alpha}$ for $\alpha \in (1, 2)$, by (5.1.5). Noticing that

$\eta(r)_t + q(r)_x$ is bounded in $W^{-1,\infty}$, and using Theorem 2.3.2, we get the proof that

$$\eta_i(r^\varepsilon(x,t))_t + q_i(r^\varepsilon(x,t))_x \text{ are compact in } W_{loc}^{-1,2}(R \times R^+), \quad (5.1.8)$$

for $i = 1, 2$, where

$$(\eta_1(r), q_1(r)) = (r - k, f(r) - f(k)) \qquad (5.1.9)$$

and

$$(\eta_2(r), q_2(r)) = (f(r) - f(k), \int_k^r (f'(s))^2 ds), \qquad (5.1.10)$$

and k is an arbitrary constant. Thus, if we consider that r is an independent variable, noticing the condition (5.0.6) on f, which is exactly the same as the proof of Theorem 3.1.1, we get the proof of $r^\varepsilon(x,t) \to l(x,t)$, almost everywhere.

Now we are going to prove the third part of Theorem 5.0.1, namely the strong convergence of $(u^\varepsilon(x,t), v^\varepsilon(x,t)) \to (u(x,t), v(x,t))$.

Since $v_0(x) \ge c_0$, then using the last part of Theorem 1.0.2, we have that $v_\varepsilon \ge c(t, c_0, \varepsilon) > 0$.

By simple calculations, we have from system (5.0.7) that

$$
\begin{aligned}
(\frac{u}{v})_t + \lambda_1(\frac{u}{v})_x &= \varepsilon(\frac{u}{v})_{xx} - \varepsilon(\frac{2u}{v^3}v_x^2 - \frac{2}{v^2}u_x v_x) \\
&= \varepsilon(\frac{u}{v})_{xx} + \frac{2\varepsilon}{v}v_x(\frac{u_x}{v} - \frac{u}{v^2}v_x) \qquad (5.1.11) \\
&= \varepsilon(\frac{u}{v})_{xx} + \frac{2\varepsilon}{v}v_x(\frac{u}{v})_x.
\end{aligned}
$$

Therefore $\dfrac{u^\varepsilon}{v^\varepsilon}$ is uniformly bounded with respect to ε when we use the maximum principle to (5.1.11).

Differentiating Equation (5.1.11) with respect to x and then multiplying the function sign θ_x to the result, where $\theta = \frac{u}{v}$, we have

$$|\theta_x|_t + (\lambda_1|\theta_x|)_x \le \varepsilon|\theta_x|_{xx} + (\frac{2\varepsilon}{v}v_x|\theta_x|)_x. \qquad (5.1.12)$$

Integrating (5.1.12) in $R \times [0, t]$, we have

$$\int_{-\infty}^{\infty} |\theta_x|(x,t)dx \le \int_{-\infty}^{\infty} |\theta_x|(x,0)dx \le M, \qquad (5.1.13)$$

which implies the pointwise convergence of a subsequence of $\dfrac{u^\varepsilon}{v^\varepsilon}$. Combining this with the result in the second part of Theorem 5.0.1, we get the pointwise convergence of a subsequence of $(u^\varepsilon, v^\varepsilon) \to (u, v)$, where the limit (u, v) is clearly a weak solution of the Cauchy problem (5.0.1), (5.0.2). Thus we complete the proof of Theorem 5.0.1. ∎

5.2   Related Results

The proof of Theorem 5.0.1 is from [Lu11]. The study of the Cauchy problem (5.0.1), (5.0.2) by using the compensated compactness method started from [Ch3], where Chen first considered the propagation and cancellation of oscillations for the weak solution. Along the second genuinely nonlinear characteristic field, the initial oscillations cannot propagate and are killed instantaneously as time evolves, but along the first linearly degenerate field, the initial oscillations can propagate. These behaviors about weak solutions of (5.0.1), (5.0.2) are coincided with that for viscosity solutions we studied in this chapter.

The first characteristic field of system (5.0.1) is of Temple type [Te], i.e., the characteristic curve in the (u, v)-plane determined by the equation $z = c$ is a straight line, where c is a constant and $z = u/v$ is the Riemann invariant corresponding to the characteristic value λ_1. In Chapter 7, we shall study another system of Temple type, called the Le Roux system.

Chapter 6

A System of Quadratic Flux

In this chapter, we consider the existence of global weak solutions for the nonlinear hyperbolic conservation laws of quadratic flux:

$$
\begin{cases}
u_t + \dfrac{1}{2}(3u^2 + v^2)_x = 0 \\
v_t + (uv)_x = 0
\end{cases}
\tag{6.0.1}
$$

with initial data

$$
(u(x,0), v(x,0)) = (u_0(x), v_0(x)) \ (v_0(x) \geq 0).
\tag{6.0.2}
$$

System (6.0.1) is the special one of the following more general systems of quadratic flux:

$$
\begin{cases}
u_t + (a_1 u^2 + a_2 uv + a_3 v^2)_x = 0 \\
v_t + (b_1 u^2 + b_2 uv + b_3 v^2)_x = 0,
\end{cases}
\tag{6.0.3}
$$

where $a_i, b_i, i = 1, 2, 3$ are all constants. System (6.0.3) can be used to approximate any given nonlinear system of two equations near the original point if we represent the nonlinear flux functions by Taylor series first, and then neglect the linear terms and the higher-order small terms.

Let F be the mapping from R^2 into R^2 defined by

$$
F : (u, v) \rightarrow (\frac{1}{2}(3u^2 + v^2), uv).
$$

Then

$$dF = \begin{pmatrix} 3u & v \\ v & u \end{pmatrix},$$

(6.0.4)

and the eigenvalues of system (6.0.1) are solutions of the following characteristic equation:

$$\lambda^2 - 4u\lambda + 3u^2 - v^2 = 0.$$

(6.0.5)

Thus two eigenvalues of system (6.0.1) are

$$\lambda_1 = 2u - s^{\frac{1}{2}}, \quad \lambda_2 = 2u + s^{\frac{1}{2}}$$

(6.0.6)

with corresponding right eigenvectors

$$r_1 = (s^{\frac{1}{2}} - u, -v)^T, \quad r_2 = (s^{\frac{1}{2}} + u, v)^T,$$

(6.0.7)

where $s = u^2 + v^2$.

The Riemann invariants of (6.0.1) are functions $w(u,v)$ and $z(u,v)$ satisfying the equations

$$w_u(s^{\frac{1}{2}} - u) - vw_v = 0 \text{ and } z_u(s^{\frac{1}{2}} + u) + vw_v = 0.$$

(6.0.8)

One solution of (6.0.8) is

$$w(u,v) = u + s^{\frac{1}{2}} \text{ and } z(u,v) = u - s^{\frac{1}{2}}.$$

(6.0.9)

By simple calculations,

$$\nabla\lambda_1 \cdot r_1 = 3(s^{\frac{1}{2}} - u), \quad \nabla\lambda_2 \cdot r_2 = 3(s^{\frac{1}{2}} + u).$$

(6.0.10)

Therefore it follows from (6.0.6) that $\lambda_1 = \lambda_2$ at point $(0,0)$, at which the strict hyperbolicity fails to hold, and from (6.0.10), the first characteristic field is linearly degenerate on $v = 0$, $u \geq 0$ and the second characteristic field is linearly degenerate on $v = 0$, $u \leq 0$.

Consider the Cauchy problem for the related parabolic system

$$\begin{cases} u_t + \frac{1}{2}(3u^2 + v^2)_x = \varepsilon u_{xx}, \\ v_t + (uv)_x = \varepsilon v_{xx} \end{cases}$$

(6.0.11)

with initial data

$$(u^\varepsilon(x,0), v^\varepsilon(x,0)) = (u_0^\varepsilon(x), v_0^\varepsilon(x)),$$

(6.0.12)

where

$$(u_0^\varepsilon(x), v_0^\varepsilon(x)) = (u_0(x), v_0(x) + \varepsilon) * G^\varepsilon \qquad (6.0.13)$$

and G^ε is a mollifier. Then

$$(u_0^\varepsilon(x), v_0^\varepsilon(x)) \in C^\infty \times C^\infty, \qquad (6.0.14)$$

$$(u_0^\varepsilon(x), v_0^\varepsilon(x)) \to (u_0(x), v_0(x)) \text{ a.e., as } \varepsilon \to 0, \qquad (6.0.15)$$

and

$$\|u_0^\varepsilon(x)\| \le M_1, \quad \varepsilon \le v_0^\varepsilon(x) \le M_1, \qquad (6.0.16)$$

for a suitable large constant M_1, which depends only on the L^∞ bound of $(u_0(x), v_0(x))$, but is independent of ε.

The main result in this chapter is given in the following theorem:

Theorem 6.0.1 *Let the initial data $(u_0(x), v_0(x))$ be bounded measurable and $v_0(x) \ge 0$. Then for fixed $\varepsilon > 0$, the viscosity solution $(u^\varepsilon(x,t), v^\varepsilon(x,t))$ of the Cauchy problem (6.0.11), (6.0.12) exists and satisfies*

$$\|u^\varepsilon(x,t)\| \le M_2, \quad 0 < c(\varepsilon,t) \le v^\varepsilon(x,t) \le M_2, \qquad (6.0.17)$$

where M_2 is a positive constant independent of ε, $c(\varepsilon,t)$ is a positive function, which could tend to zero as ε tends to zero or t tends to infinity.

Moreover, there exists a subsequence (still labelled) $(u^\varepsilon(x,t), v^\varepsilon(x,t))$ such that

$$(u^\varepsilon(x,t), v^\varepsilon(x,t)) \to (u(x,t), v(x,t)), \quad a.e. \text{ on } \Omega, \qquad (6.0.18)$$

where $\Omega \subset R \times R^+$ is any bounded open set, and $(u(x,t), v(x,t))$ is a weak solution of the Cauchy problem (6.0.1), (6.0.2).

The above theorem includes two parts: one is the existence of viscosity solutions and related estimates (6.0.17), whose proof is given in Section 6.1; the other part is about the strong convergence (6.0.18) of a subsequence of $(u^\varepsilon(x,t), v^\varepsilon(x,t))$, whose proof is given in Sections 6.2-6.4.

6.1   Existence of Viscosity Solutions

To prove the existence of the viscosity solutions in Theorem 6.0.1, it is sufficient to get the *a priori* L^∞ estimate of $(u^\varepsilon(x,t), v^\varepsilon(x,t))$ if we use the general framework given in Theorem 1.0.2.

By simple calculations,

$$w_{uu} = \frac{v^2}{s^3}, \quad w_{uv} = -\frac{uv}{s^3}, \quad w_{vv} = \frac{u^2}{s^3},$$

and

$$z_{uu} = -\frac{v^2}{s^3}, \quad z_{uv} = \frac{uv}{s^3}, \quad z_{vv} = -\frac{u^2}{s^3}.$$

Then $w(u,v)$ and $-z(u,v)$ both are convex functions. Using Theorem 4.2.1, we have

$$\Sigma_1 = \{(u,v) : w(u,v) \le M, \ z(u,v) \ge -M\} \tag{6.1.1}$$

is an invariant region for a suitable large constant M (see Figure 6.1).

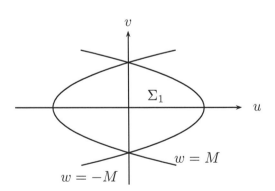

FIGURE 6.1

Thus we have the uniform L^∞ estimates

$$\|u^\varepsilon(x,t)\| \le M_2, \quad \|v^\varepsilon(x,t)\| \le M_2$$

and hence the existence of the viscosity solutions. The positive lower bound of $v^\varepsilon \ge c(\varepsilon, t) > 0$ follows from the last part of Theorem 1.0.2.

6.2 Entropy-Entropy Flux Pairs of Lax Type

A pair $(\bar{\eta}(u,v), \bar{q}(u,v))$ of real-valued functions is a pair of entropy-entropy flux of system (6.0.1) if they satisfy the following system of two equations:

$$\nabla\bar{\eta}(u,v) \cdot dF(u,v) = \nabla\bar{q}(u,v) \tag{6.2.1}$$

or equivalently

$$\bar{q}_u = 3u\bar{\eta}_u + v\bar{\eta}_v, \quad \bar{q}_v = v\bar{\eta}_u + u\bar{\eta}_v. \tag{6.2.2}$$

Eliminating $\bar{q}$ from (6.2.2), we get

$$v(\bar{\eta}_{vv} - \bar{\eta}_{uu}) + 2u\bar{\eta}_{uv} = 0. \tag{6.2.3}$$

By the definition of Riemann invariants in Chapter 4, we can easily prove that w, z satisfy

$$\nabla w(u,v) \cdot dF(u,v) = \lambda_2 \nabla w(u,v), \quad \nabla z(u,v) \cdot dF(u,v) = \lambda_1 \nabla z(u,v). \tag{6.2.4}$$

If we consider that the entropy-entropy flux pair of system (6.0.1) are functions of variables w, z: $(\bar{\eta}, \bar{q}) = (\bar{\eta}(w,z), \bar{q}(w,z))$, then

$$\bar{q}(w,z)_w = \lambda_2\bar{\eta}(w,z)_w, \quad \bar{q}(w,z)_z = \lambda_1\bar{\eta}(w,z)_z. \tag{6.2.5}$$

Eliminating $\bar{q}$ from (6.2.4), we have

$$(\lambda_2 - \lambda_1)\bar{\eta}(w,z)_{wz} + \lambda_{2z}\bar{\eta}(w,z)_w - \lambda_{1w}\bar{\eta}(w,z)_z = 0. \tag{6.2.6}$$

Noticing that

$$\lambda_2 = 2u + s^{\frac{1}{2}} = w + z + \frac{w-z}{2} = \frac{3w+z}{2}$$

and

$$\lambda_1 = 2u - s^{\frac{1}{2}} = w + z - \frac{w-z}{2} = \frac{w+3z}{2},$$

we get the other entropy equation of system (6.0.1) in the following form:

$$\bar{\eta}(w,z)_{wz} + \frac{1}{2(w-z)}(\bar{\eta}(w,z)_w - \bar{\eta}(w,z)_z) = 0. \tag{6.2.7}$$

Now we make a transformation of variables from (u, v) to (u, s), that is

$$\bar{\eta}(u, v) = \eta(u, s), \quad \bar{q}(u, v) = q(u, s).$$

By simple calculations, we have

$$\bar{\eta}_u = \eta_u + 2u\eta_s, \quad \bar{\eta}_v = 2v\eta_s, \quad \bar{\eta}_{vv} = 4v^2\eta_{ss} + 2\eta_s$$

and

$$\bar{\eta}_{uv} = 4uv\eta_{ss} + 2v\eta_{us}, \quad \bar{\eta}_{uu} = 4u^2\eta_{ss} + 2\eta_s + 4u\eta_{us} + \eta_{uu}.$$

Then the entropy equation (6.2.7) is changed to the following simple equation:

$$\eta_{ss} = \frac{1}{4s}\eta_{uu}. \tag{6.2.8}$$

It follows from (6.2.2) that

$$\begin{cases} 2uq_s + q_u = 3u(2u\eta_s + \eta_u) + 2v^2\eta_s, \\ 2vq_s = v(2u\eta_s + \eta_u) + 2uv\eta_s \end{cases} \tag{6.2.9}$$

and hence the entropy flux q corresponding to the entropy η satisfies

$$q_u = 2u\eta_u + 2s\eta_s. \tag{6.2.10}$$

If k denotes a constant, then the function $\eta = h(s)e^{ku}$ solves (6.2.8) provided that

$$h''(s) - \frac{k^2}{4s}h(s) = 0. \tag{6.2.11}$$

Let $a(s) = s^{\frac{1}{4}}, r = ks^{\frac{1}{2}}, h(s) = a(s)\phi(r)$. Then a simple calculation shows that

$$\phi''(r) - (1 + \frac{3}{4r^2})\phi(r) = 0, \tag{6.2.12}$$

which is the standard Fuchsian equation. We can look for a series solution of (6.2.12) with the following form:

$$\phi_1(r) = r^{\frac{3}{2}} \sum_{n=0}^{\infty} c_n r^{2n} = r^{\frac{3}{2}} g(r). \tag{6.2.13}$$

Then the coefficients c_n must satisfy

$$c_n = \frac{c_{n-1}}{(\frac{3}{2} + 2n)(\frac{1}{2} + 2n) - \frac{3}{4}}, \quad \text{for } n \geq 1 \tag{6.2.14}$$

and c_0 could be any positive constant.

Let another independent solution ϕ_2 of (6.2.12) satisfy $\phi_2 = \phi_1 P$. Then P solves

$$P'' \phi_1 + 2\phi_1' P' = 0. \tag{6.2.15}$$

Thus

$$P' = -(\phi_1)^{-2} = -(r^3 g^2(r))^{-1},$$

and one function P is given by

$$P = \int_r^\infty (r^3 g^2(r))^{-1} dr. \tag{6.2.16}$$

Therefore

$$\phi_2(r) = r^{\frac{3}{2}} g(r) \int_r^\infty (r^3 g^2(r))^{-1} dr. \tag{6.2.17}$$

If $\eta_k = a(s)\phi(r)e^{ku}$, then we have from (6.2.10) that

$$(q_k)_u = 2ku\eta_k + (\frac{1}{2} + r\frac{\phi'(r)}{\phi(r)})\eta_k \tag{6.2.18}$$

and hence, one entropy flux q_k corresponding to η_k is

$$q_k = \eta_k(2u + s^{\frac{1}{2}} + \frac{r}{k}(\frac{\phi'(r)}{\phi(r)} - 1) - \frac{3}{2k}). \tag{6.2.19}$$

Let $\eta_{-k} = a(s)\phi(r)e^{-ku}$, then by (6.2.10), one entropy flux q_{-k} corresponding to η_{-k} is

$$q_{-k} = \eta_{-k}(2u - s^{\frac{1}{2}} - \frac{r}{k}(\frac{\phi'(r)}{\phi(r)} - 1) + \frac{3}{2k}). \tag{6.2.20}$$

About the solutions ϕ_1, ϕ_2 of the Fuchsian equation

$$\phi'' - (1 + \frac{c}{r^2})\phi = 0, \tag{6.2.21}$$

where c is a constant, we have the following lemma:

Lemma 6.2.1 *If $\phi_1(r) > 0$, $\phi_1'(r) > 0$ for $r > 0$, then*

$$\frac{\phi_1'(r)}{\phi_1(r)} = 1 + O(\frac{1}{r^2}), \quad c_1\phi_1(r)e^{-r} = 1 + O(\frac{1}{r}) \tag{6.2.22}$$

as r approaches infinity;
   if $\phi_2(r) > 0, \phi_2'(r) < 0$ for $r > 0$, then

$$\frac{\phi_2'(r)}{\phi_2(r)} = -1 + O(\frac{1}{r^2}), \quad c_2\phi_2(r)e^{r} = 1 + O(\frac{1}{r}) \tag{6.2.23}$$

as r approaches infinity, where c_1, c_2 are two suitable, positive constants.

Proof. Since $\phi'' = (1 + \frac{c}{r^2})\phi$, if $\phi_1(r) > 0, \phi_1'(r) > 0$, then $\phi_1(r), \phi_1'(r)$ both tend to infinite as r approaches infinity; if $\phi_2(r) > 0, \phi_2'(r) < 0$, then $\phi_2(r), \phi_2'(r)$ both go to zero as r goes to infinity. Thus $\dfrac{\phi_1'(r)}{\phi_1(r)}$ has the form $\dfrac{\infty}{\infty}$ and $\dfrac{\phi_2'(r)}{\phi_2(r)}$ has the form $\dfrac{0}{0}$ as r goes to infinity. Therefore we have by the Vol'pert theorem,

$$\lim_{r\to\infty} \frac{\phi'(r)}{\phi(r)} \lim_{r\to\infty} \frac{\phi''(r)}{\phi'(r)} \lim_{r\to\infty} \frac{(1 + \frac{c}{r^2})\phi}{\phi'(r)}. \tag{6.2.24}$$

Then

$$\lim_{r\to\infty} (\frac{\phi'(r)}{\phi(r)})^2 = 1. \tag{6.2.25}$$

So we have

$$\lim_{r\to\infty} \frac{\phi_1'(r)}{\phi_1(r)} = 1, \quad \lim_{r\to\infty} \frac{\phi_2'(r)}{\phi_2(r)} = -1. \tag{6.2.26}$$

   Using (6.2.21), we have

$$(\frac{\phi'(r)}{\phi(r)})' + (\frac{\phi'(r)}{\phi(r)})^2 = 1 + \frac{c}{r^2}. \tag{6.2.27}$$

Let $y = \dfrac{\phi'(r)}{\phi(r)} - 1$. Then y satisfies

$$y' + (1 + \frac{\phi'(r)}{\phi(r)})y = \frac{c}{r^2}. \tag{6.2.28}$$

Integrating Equation (6.2.28) from d to r (d is a positive constant), we have

$$e^{\int_d^r (\frac{\phi_1'(t)}{\phi_1(t)}+1)dt} y(r) - y(d) = \int_d^r e^{\int_d^\tau (\frac{\phi_1'(t)}{\phi_1(t)}+1)dt} \frac{c}{\tau^2} d\tau, \qquad (6.2.29)$$

which implies that

$$\frac{\phi_1(r)}{\phi_1(d)} e^{r-d} y(r) - y(d) = \int_d^r \frac{\phi_1(\tau)}{\phi_1(d)} e^{\tau-d} \frac{c}{\tau^2} d\tau,$$

and hence

$$y(r) = \frac{\phi_1(d)}{\phi_1(r)} e^{d-r} y(d) + \int_d^r \frac{\phi_1(\tau)}{\phi_1(r)} e^{\tau-r} \frac{c}{\tau^2} d\tau. \qquad (6.2.30)$$

Let $d = \frac{1}{2}r$. Then clearly $\phi_1(d) < \phi_1(r), \phi_1(\tau) < \phi_1(r)$ since $\phi_1' > 0$.
 Since y is bounded from (6.2.26), we have from (6.2.30) that

$$|y(r)| \leq M e^{-\frac{1}{2}r} + \frac{4|c|}{r^2}(1 - e^{-\frac{1}{2}r}), \qquad (6.2.31)$$

where M is the bound of y. Therefore

$$y(r) = O(\frac{1}{r^2}) \qquad (6.2.32)$$

as r approaches infinity. The first part in (6.2.22) is proved.
 Similarly let $z = \frac{\phi_2'(r)}{\phi_2(r)} + 1$. We have from (6.2.27) that

$$z' + (\frac{\phi'(r)}{\phi(r)} - 1)z = \frac{c}{r^2}. \qquad (6.2.33)$$

Integrating (6.2.33) from r to d, we have

$$z(r) = \frac{\phi_2(d)}{\phi_2(r)} e^{r-d} z(d) - \int_r^d \frac{\phi_2(\tau)}{\phi_2(r)} e^{r-\tau} \frac{c}{\tau^2} d\tau. \qquad (6.2.34)$$

Let $d = 2r$ in (6.2.34). Then clearly $\phi_2(d) < \phi_2(r), \phi_2(\tau) < \phi_2(r)$ since $\phi_2' < 0$. Thus we have

$$|z(r)| \leq M e^{-r} + \frac{|c|}{r^2}(1 - e^{-r}), \qquad (6.2.35)$$

which implies that

$$z(r) = O(\frac{1}{r^2}) \tag{6.2.36}$$

as r approaches infinity. The first part in (6.2.23) is proved.

Now we prove the remaining estimates in (6.2.22) and (6.2.23).

Let $\phi_1(r) = a(r)e^r$. Then

$$\frac{a'(r)}{a(r)} = y(r) = O(\frac{1}{r^2}) \tag{6.2.37}$$

for r large. Thus

$$a(r) = a(r_1)e^{\int_{r_1}^r y(r)dr}, \tag{6.2.38}$$

for any fixed $r_1 > 0$.

Since $y(r) = O(\frac{1}{r^2})$ as r approaches infinity, the limit of $\lim_{r\to\infty} a(r)$ exists. Then (6.2.38) gives

$$\begin{aligned} a(r) &= a(\infty)e^{-\int_r^\infty y(r)dr} \\ &= a(\infty) + a(\infty)(e^{-\int_r^\infty y(r)dr} - 1). \end{aligned} \tag{6.2.39}$$

Since $\int_r^\infty y(r)dr = O(\frac{1}{r})$ as r approaches infinity, applying the Taylor expansion to $e^{-\int_r^\infty y(r)dr} - 1$ gives

$$a(r) = a(\infty) + O(\frac{1}{r}) \tag{6.2.40}$$

as r approaches infinity.

Similarly let $\phi_2(r) = b(r)e^{-r}$. Then

$$\frac{b'(r)}{b(r)} = z(r). \tag{6.2.41}$$

Since $z(r) = O(\frac{1}{r^2})$ as r approaches infinity, we have

$$b(r) = b(\infty) + O(\frac{1}{r}) \tag{6.2.42}$$

as r approaches infinity. Therefore Lemma 6.2.1 is proved. ∎

It is clear that $\phi_1(r), \phi_2(r)$ given in (6.2.13) and (6.2.17) satisfy that $\phi_1(r) > 0, \phi_1'(r) > 0$ and $\phi_2(r) > 0$ for all $s > 0$. The strict positivity of $\phi_2''(r)$ gives $\phi_2'(r) < 0$ as $s > 0$ because $\lim_{r \to \infty} \phi_2(r) = 0$, $\lim_{r \to \infty} \phi_2'(r) = 0$.

Applying the estimates in (6.2.22)-(6.2.23) to $\phi_1(r), \phi_2(r)$, we have

$$\eta_k^1 = a(s)\phi_1(r)e^{ku}$$

$$= e^{kw}(a(s) + O(\tfrac{1}{r})) = e^{kw}(a(s) + O(\tfrac{1}{k})) \tag{6.2.43}$$

on any compact subset of $s > 0$ since $r = ks^{\frac{1}{2}}$;

$$q_k^1 = \eta_k^1(2u + s^{\frac{1}{2}} + \frac{r}{k}(\frac{\phi_1'(r)}{\phi_1(r)} - 1) - \frac{3}{2k})$$

$$= \eta_k^1(\lambda_2 + O(\frac{1}{k})) \tag{6.2.44}$$

on $s \geq 0$ by the fact that factor $r(\frac{\phi_1'(r)}{\phi_1(r)} - 1)$ is uniformly bounded from (6.2.31). Furthermore

$$q_k^1 = \eta_k^1(\lambda_2 - \frac{3}{2k} + O(\frac{1}{k^2})) \tag{6.2.45}$$

on any compact subset of $s > 0$;

$$\eta_{-k}^1 = a(s)\phi_1(r)e^{-ku} = e^{-kz}(a(s) + O(\frac{1}{k})) \tag{6.2.46}$$

on any compact subset of $s > 0$;

$$q_{-k}^1 = \eta_{-k}^1(\lambda_1 - \frac{r}{k}(\frac{\phi_1'(r)}{\phi_1(r)} - 1) + \frac{3}{2k})$$

$$= \eta_{-k}^1(\lambda_1 + O(\frac{1}{k})) \tag{6.2.47}$$

on $s \geq 0$ and

$$q_{-k}^1 = \eta_{-k}^1(\lambda_1 + \frac{3}{2k} + O(\frac{1}{k^2})) \tag{6.2.48}$$

on any compact subset of $s > 0$;

$$\eta_k^2 = a(s)\phi_2(r)e^{ku} = e^{kz}(a(s) + O(\frac{1}{k})) \tag{6.2.49}$$

on any compact subset of $s > 0$;

$$q_k^2 = \eta_k^2(\lambda_1 + \frac{r}{k}(\frac{\phi_2'(r)}{\phi_2(r)} + 1) - \frac{3}{2k})$$

$$= \eta_k^2(\lambda_1 + O(\frac{1}{k}))$$

(6.2.50)

on $s \geq 0$ by the fact that factor $r(\frac{\phi_1'(r)}{\phi_1(r)} + 1)$ is uniformly bounded from (6.2.35). Furthermore

$$q_k^2 = \eta_k^2(\lambda_1 - \frac{3}{2k} + O(\frac{1}{k^2}))$$

(6.2.51)

on any compact subset of $s > 0$;

$$\eta_{-k}^2 = a(s)\phi_2(r)e^{-ku} = e^{-kw}(a(s) + O(\frac{1}{k}))$$

(6.2.52)

on any compact subset of $s > 0$;

$$q_{-k}^2 = \eta_{-k}^2(\lambda_2 - \frac{r}{k}(\frac{\phi_2'(r)}{\phi_2(r)} + 1) + \frac{3}{2k})$$

$$= \eta_{-k}^2(\lambda_2 + O(\frac{1}{k}))$$

(6.2.53)

on $s \geq 0$ and

$$q_{-k}^2 = \eta_{-k}^2(\lambda_2 + \frac{3}{2k} + O(\frac{1}{k^2}))$$

(6.2.54)

on any compact subset of $s > 0$.

The above estimates about the entropy-entropy flux pairs will be used to prove the Young measure ν, corresponding to the sequence of viscosity solutions, is a Dirac measure in Section 6.4.

6.3   Compactness of $\eta_t + q_x$ in H_{loc}^{-1}

In this section, we mainly prove the following theorem:

Theorem 6.3.1 *For the entropy-entropy flux pairs (η, q) of Lax type constructed in Section 6.2,*

$$\eta(u^\varepsilon, v^\varepsilon)_t + q(u^\varepsilon, v^\varepsilon)_x$$

is compact in $H_{loc}^{-1}(R \times R^+)$ with respect to the approximated solutions $(u^\varepsilon, v^\varepsilon)$ constructed by the viscosity method.

Proof. For simplicity, we drop the superscript ε in the viscosity solutions $(u^\varepsilon, v^\varepsilon)$.

It is obvious that system (6.0.1) has a strictly convex entropy $\eta^\star = \frac{u^2+v^2}{2}$ and the corresponding entropy flux $q^\star = u^3 + uv^2$.

Multiplying the first equation in (6.0.11) by u and the second by v, then adding the result, we have

$$\eta_t^\star + q_x^\star = \varepsilon\eta_{xx}^\star - \varepsilon(\eta_{uu}^\star u_x^2 + 2\eta_{uv}^\star u_x v_x + \eta_{vv}^\star v_x^2) = \varepsilon\eta_{xx}^\star - \varepsilon(u_x^2 + v_x^2). \tag{6.3.1}$$

Using the same technique as given in obtaining (5.1.5), we have

$$\varepsilon(u_x^\varepsilon)^2 \text{ and } \varepsilon(v_x^\varepsilon)^2 \text{ are bounded in } L_{loc}^1(R \times R^+). \tag{6.3.2}$$

The first class of entropy-entropy flux pair of Lax type related to the function ϕ_1 constructed in Section 6.2, denoted by $\eta_{\pm k}^1$, are clearly smooth functions of (u, v). In fact

$$\eta_{\pm k}^1 = k^{\frac{3}{2}}s \sum_{n=0}^{\infty} c_n (k^2 s)^n e^{\pm ku}. \tag{6.3.3}$$

So Theorem 6.3.1 can be easily proved for the first class of entropy-entropy flux pair of Lax type.

However the second order derivatives of the second class of entropy-entropy flux pair of Lax type related to the function ϕ_2 constructed in Section 6.2, denoted by $\eta_{\pm k}^2$, are singular at the point $(u, v) = (0, 0)$. In fact

$$\eta_{\pm k}^2 = k^{-\frac{1}{2}}e^{\pm ku}r^2 g(r) \int_r^\infty (r^3 g^2(r))^{-1} dr, \tag{6.3.4}$$

where

$$\int_r^\infty (r^3 g^2(r))^{-1} dr = O(\frac{1}{r^2}), \tag{6.3.5}$$

as r approaches zero, and hence for any fixed $k > 0$, $\eta_{\pm k}^2$ and $q_{\pm k}^2$ are uniformly bounded from (6.2.49)-(6.2.53).

Moreover,

$$\begin{aligned}
\eta_{\pm k}^2 &= k^{-\frac{1}{2}}e^{\pm ku}r^2 g(r) \int_r^\infty (r^3 g^2(r))^{-1} dr \\
&= k^{-\frac{1}{2}}e^{\pm ku}\left(\frac{1}{2g(r)} - r^2 g(r) \int_r^\infty \frac{g'(r)}{r^2 g^3(r)} dr\right),
\end{aligned} \tag{6.3.6}$$

where

$$g'(r) = \sum_{n=1}^{\infty} 2nc_n r^{2n-1} \leq \sum_{n=1}^{\infty} c_{n-1} r^{2n-1} = rg(r), \qquad (6.3.7)$$

thus

$$r^2 g(r) \int_r^{\infty} \frac{g'(r)}{r^2 g^3(r)} dr = O(r^2 \log r) \qquad (6.3.8)$$

as r approaches zero. This implies that for any fixed $k > 0$, the first order derivatives of $\eta_{\pm k}^2$ are uniformly bounded. It is clear that the first part $I_1 = k^{-\frac{1}{2}} e^{\pm ku} \frac{1}{2g(r)}$ in $\eta_{\pm k}^2$ are smooth; its second order derivatives are bounded. But the second part in $\eta_{\pm k}^2$ can be written as $r^2 I_2$, where

$$I_2 = -k^{-\frac{1}{2}} e^{\pm ku} g(r) \int_r^{\infty} \frac{g'(r)}{r^2 g^3(r)} dr, \qquad (6.3.9)$$

its second order derivatives are singular at the point $(0,0)$. In fact, from (6.3.8), all derivatives of second order of function $r^2 I_2$ are bounded except the term $\left((r^2)_{uu} + (r^2)_{vv}\right) I_2 = 2I_2$, which is positive.

Therefore, multiplying system (6.0.11) by $\nabla \eta_{\pm k}^2$, we have

$$(\eta_{\pm k}^2)_t + (q_{\pm k}^2)_x$$

$$\varepsilon(\eta_{\pm k}^2)_{xx} - \varepsilon\left((\eta_{\pm k}^2)_{uu} u_x^2 + 2(\eta_{\pm k}^2)_{uv} u_x v_x + (\eta_{\pm k}^2)_{vv} v_x^2\right)$$
$$= \varepsilon(\eta_{\pm k}^2)_{xx} - \varepsilon\left(A(u,v) u_x^2 + B(u,v) u_x v_x + C(u,v) v_x^2\right) \qquad (6.3.10)$$

$$-2\varepsilon I_2(u_x^2 + v_x^2),$$

where $A(u,v), B(u,v)$ and $C(u,v)$ are the regular derivatives of second order of $\eta_{\pm k}^2$.

Let $K \subset R \times R^+$ be an arbitrary compact set and choose $\phi \in C_0^{\infty}(R \times R^+)$ such that $\phi_K = 1, 0 \leq \phi \leq 1$ and $S = \text{supp}\,\phi$.

Multiplying Equation (6.3.10) by ϕ and integrating over $R \times R^+$, we obtain

$$\varepsilon \int_0^{\infty} \int_{-\infty}^{\infty} 2\varepsilon I_2(u_x^2 + v_x^2)\phi dx dt$$

$$= -\varepsilon\left(A(u,v) u_x^2 + B(u,v) u_x v_x + C(u,v) v_x^2\right)\phi \qquad (6.3.11)$$

$$+\eta_{\pm k}^2 \phi_t + q_{\pm k}^2 \phi_x \varepsilon \eta_{\pm k}^2 \phi_{xx} \leq M(\phi),$$

where the last inequality follows from the boundedness of viscosity solutions, the local boundedness in L^1_{loc} in (6.3.2) of the regular part $A(u,v)u_x^2 + B(u,v)u_x v_x + C(u,v)v_x^2$.

Considering (6.3.10) again, we see that the part

$$\varepsilon\big((\eta_{\pm k}^2)_{uu}u_x^2 + 2(\eta_{\pm k}^2)_{uv}u_x v_x + (\eta_{\pm k}^2)_{vv}v_x^2\big)$$

is bounded in L^1_{loc} and hence, compact in $W^{-1,\alpha}_{loc}$ for a constant $\alpha \in (1,2)$. The part $\varepsilon(\eta_{\pm k}^2)_{xx}$ is clearly compact in $W^{-1,2}_{loc}$ because the boundedness of derivatives of the first order of $\eta_{\pm k}^2$, and the L^1_{loc} estimates (6.3.2) for u_x^2 and v_x^2. Noticing the boundedness of

$$(\eta_{\pm k}^2)_t + (q_{\pm k}^2)_x$$

in $W^{-1,\infty}$, we get the proof of Theorem 6.3.1 by Theorem 2.3.2. ∎

6.4 Reduction of ν

In this section, we shall prove that the family of positive measures $\nu_{x,t}$, determined by the sequence of viscosity solutions $(u^\varepsilon(x,t), v^\varepsilon(x,t))$ of the Cauchy problem (6.0.11), (6.0.12), must be Dirac measures. Then using Theorem 2.2.3, we get the proof of (6.0.18), which implies that the function $(u(x,t), v(x,t))$ of support set points of these Dirac measures is a weak solution of the Cauchy problem (6.0.1), (6.0.2).

Since the viscosity solutions $(u^\varepsilon(x,t), v^\varepsilon(x,t))$ of the Cauchy problem (6.0.11), (6.0.12) are uniformly bounded in L^∞ space, by Theorem 2.2.1, we consider the family of compactly supported probability measures $\nu_{x,t}$. Without any loss of generality we may fix $(x,t) \in R \times R^+$ and consider only one measure ν.

For any entropy-entropy flux pairs (η_i, q_i), $i = 1, 2$, of system (6.0.1), satisfying the compactness of $\eta(u^\varepsilon, v^\varepsilon)_t + q(u^\varepsilon, v^\varepsilon)_x$ in $H^{-1}_{loc}(R \times R^+)$, we have from Theorem 2.1.4 that

$$\overline{\eta_1(u^\varepsilon, v^\varepsilon)} \cdot \overline{q_2(u^\varepsilon, v^\varepsilon)} - \overline{\eta_2(u^\varepsilon, v^\varepsilon)} \cdot \overline{q_1(u^\varepsilon, v^\varepsilon)}$$

$$= \overline{\eta_1(u^\varepsilon, v^\varepsilon)q_2(u^\varepsilon, v^\varepsilon) - \eta_2(u^\varepsilon, v^\varepsilon)q_1(u^\varepsilon, v^\varepsilon)}. \tag{6.4.1}$$

Then using Theorem 2.2.1, we have the following measure equation:

$$< \nu, \eta_1 >< \nu, q_2 > - < \nu, \eta_2 >< \nu, q_1 >< \nu, \eta_1 q_2 - \eta_2 q_1 > . \tag{6.4.2}$$

Let Q denote the smallest characteristic rectangle:

$$Q = \{(u, v) : w_- \leq w \leq w_+, \ z_- \leq z \leq z_+, \ v \geq 0\}.$$

We now prove that the support set of ν is either contained in the point $(0, 0)$ or in another point $(w^\star, z^\star)$.

We assume that supp ν is not the unique point $(0, 0)$. Then $< \nu, \eta_k^1 > > 0$ and $< \nu, \eta_{-k}^2 > > 0$, where η_k^1, η_{-k}^2 are given by (6.2.43) and (6.2.52).

We introduce two new probability measures μ_k^+ and μ_k^- on Q, defined by

$$< \mu_k^+, h >=< \nu, h\eta_k^1 > / < \nu, \eta_k^1 > \qquad (6.4.3)$$

and

$$< \mu_k^-, h >=< \nu, h\eta_{-k}^2 > / < \nu, \eta_{-k}^2 >, \qquad (6.4.4)$$

where $h = h(u, v)$ denotes an arbitrary continuous function. Clearly μ_k^+ and μ_k^- both are uniformly bounded with respect to k. Then as a consequence of weak-star compactness, there exist probability measures $\mu^\pm$ on Q such that

$$< \mu^\pm, h >= \lim_{k \to \infty} < \mu_k^\pm, h > \qquad (6.4.5)$$

after the selection of an appropriate subsequence.

Moreover, the measures μ^+, μ^- are respectively concentrated on the boundary sections of Q associated with w, i.e.

$$\text{supp } \mu^+ = Q \bigcap \{(u, v) : w = w_+\} \qquad (6.4.6)$$

and

$$\text{supp } \mu^- = Q \bigcap \{(u, v) : w = w_-\}. \qquad (6.4.7)$$

In fact, for any function $h(w, z) \in C_0(Q)$, satisfying

$$\text{supp } h(w, z) \subset Q \bigcap \{w \leq w_0\},$$

where $w_0 < w_+$ is any number, as $k \to \infty$, we have

$$\frac{| < \nu, h\eta_k^1 > |}{| < \nu, \eta_k^1 > |} = \frac{| < \nu, he^{kw}(a(s) + O(\frac{1}{k})) > |}{| < \nu, e^{kw}(a(s) + O(\frac{1}{k})) > |} \qquad (6.4.8)$$

$$\leq \frac{c_1 e^{k(w_0+\delta)}}{c_2 e^{k(w_+-\delta)}} = \frac{c_1}{c_2} e^{k(w_0+2\delta-w_+)} \to 0,$$

where c_1, c_2 are two suitable positive constants and $\delta > 0$ satisfies $2\delta < w_+ - w_0$, since Q is the smallest characteristic rectangle of ν. Thus we get the proof of (6.4.6). Similarly we can prove (6.4.7).

Let $(\eta_1, q_1) = (\eta_k^1, q_k^1)$ in (6.4.2). We have

$$< \nu, q_2 > - < \nu, \eta_2 > \frac{< \nu, q_k^1 >}{< \nu, \eta_k^1 >} = \frac{< \nu, \eta_k^1 q_2 - \eta_2 q_k^1 >}{< \nu, \eta_k^1 >}. \qquad (6.4.9)$$

Noticing the estimate (6.2.45) between η_k^1 and q_k^1, and letting $k \to \infty$ in (6.4.9), we have

$$< \nu, q_2 > - < \nu, \eta_2 >< \mu^+, \lambda_2 >=< \mu^+, q_2 - \lambda_2 \eta_2 > . \qquad (6.4.10)$$

Similarly, let $(\eta_1, q_1) = (\eta_{-k}^2, q_{-k}^2)$ in (6.4.2) and use the estimate (6.2.54) between η_{-k}^2 and q_{-k}^1. We have

$$< \nu, q_2 > - < \nu, \eta_2 >< \mu^-, \lambda_2 >=< \mu^-, q_2 - \lambda_2 \eta_2 > . \qquad (6.4.11)$$

Let $(\eta_1, q_1) = (\eta_k^1, q_k^1)$ and $(\eta_2, q_2) = (\eta_{-k}^2, q_{-k}^2)$ in (6.4.2). We have

$$\frac{< \nu, q_{-k}^2 >}{< \nu, \eta_{-k}^2 >} - \frac{< \nu, q_k^1 >}{< \nu, \eta_k^1 >} = \frac{< \nu, \eta_k^1 q_{-k}^2 - \eta_{-k}^2 q_k^1 >}{< \nu, \eta_{-k}^2 >< \nu, \eta_k^1 >}. \qquad (6.4.12)$$

We now prove that $w_- = w_+$. If not, choose $\delta_0 > 0$ such that $2\delta_0 < w_+ - w_-$. Then

$$< \nu, \eta_{-k}^2 > \geq c_1 e^{-k(w_- + \delta_0)}, \qquad < \nu, \eta_k^1 > \geq c_2 e^{k(w_+ - \delta_0)} \qquad (6.4.13)$$

for two suitable positive constants c_1, c_2 and hence, the right-hand side of (6.4.12) satisfies

$$\frac{< \nu, \eta_k^1 q_{-k}^2 - \eta_{-k}^2 q_k^1 >}{< \nu, \eta_{-k}^2 >< \nu, \eta_k^1 >} = O(\frac{1}{k})e^{-k(w_+ - w_- - 2\delta_0)} \to 0, \qquad \text{as } k \to \infty,$$
$$(6.4.14)$$

resulting from the estimates given by (6.2.43), (6.2.44), (6.2.52) and (6.2.53).

Letting $k \to \infty$ in (6.4.12), we have $< \nu^+, \lambda_2 >=< \nu^-, \lambda_2 >$. Combining this with (6.4.10)-(6.4.11) gives the following relation:

$$< \mu^+, q - \lambda_2 \eta >< \mu^-, q - \lambda_2 \eta > \qquad (6.4.15)$$

for any (η, q) satisfying that $\eta_t + q_x$ is compact in $H^{-1}_{loc}(R \times R^+)$.

Let (η, q) in (6.4.15) be (η^2_{-k}, q^2_{-k}). If $w_+ - w_- > 2\delta_0$, we get from the left-hand side of (6.4.15) that

$$| < \mu^+, q - \lambda_2\eta > | \geq \frac{c_1}{k}e^{k(w_+ - \delta_0)} \qquad (6.4.16)$$

and from the right-hand side

$$| < \mu^-, q - \lambda_2\eta > | \leq \frac{c_2}{k}e^{-k(w_- + \delta_0)} \qquad (6.4.17)$$

for two positive constants c_1, c_2. This is impossible and hence, w_+ must equal w_-.

Similar to the above proof, we can use entropy-entropy flux pairs $(\eta^2_k, q^2_k), (\eta^1_{-k}, q^1_{-k})$ constructed in Section 6.2 to prove $z_+ = z_-$. Thus the support set of ν is either $(0, 0)$ or another point $(w^\star, z^\star)$. This completes the proof of Theorem 6.0.1. ∎

6.5   Related Results

The Riemann solutions for general nonstrictly hyperbolic conservation laws of quadratic flux (6.0.3) were constructed by Issacson, Marchesin, Paes-Leme, Plohr, Schaeffer, Shearer, Temple and others (cf. [IMPT, IT, SSMP]).

The interactions of elementary waves for systems

$$\begin{cases} u_t + \frac{1}{2}(au^2 + v^2)_x = 0 \\ v_t + (uv)_x = 0 \end{cases} \qquad (6.5.1)$$

were studied for the case of $a > 2$ by Lu and Wang [LuW].

The global existence of weak solution to the Cauchy problem (6.0.1), (6.0.2) was obtained in [Lu4]. The proof in this chapter is from [Lu4] and [Lu8]. The entropies constructed in Section 6.2 touch the singular point (0,0) and hence, refine the proof.

In [Ka], a different proof was given by Kan independently to the Cauchy problem (6.0.1), (6.0.2). Kan's proof is based on constructing entropies $\eta(u, v)$ for system (6.0.1) away from the singular point $(u, v) = (0, 0)$. The ideas in [Ka] were applied to study L^∞ solutions for more general systems of quadratic flux in the form (6.0.3) by Chen and Kan (cf. [CK]).

Chapter 7

Le Roux System

In this chapter, we shall use the method given in Chapter 6 to study the existence of global weak solutions for the following nonlinear hyperbolic conservation laws:

$$\begin{cases} u_t + (u^2 + v)_x = 0, \\ v_t + (uv)_x = 0 \end{cases} \tag{7.0.1}$$

with initial data

$$(u(x,0), v(x,0)) = (u_0(x), v_0(x)) \ (v_0(x) \geq 0). \tag{7.0.2}$$

System (7.0.1) was first derived by Le Roux in [Le] as a mathematical model, and so is called the Le Roux system.

Let F be the mapping from R^2 into R^2 defined by

$$F : (u, v) \rightarrow (u^2 + v, uv).$$

Then

$$dF = \begin{pmatrix} 2u & 1 \\ v & u \end{pmatrix}, \tag{7.0.3}$$

and the eigenvalues of system (7.0.1) are solutions of the following characteristic equation:

$$\lambda^2 - 3u\lambda + 2u^2 - v = 0. \tag{7.0.4}$$

Two roots of Equation (7.0.4) are

$$\lambda_1 = \frac{3u}{2} - \frac{D}{2}, \quad \lambda_2 = \frac{3u}{2} + \frac{D}{2}, \tag{7.0.5}$$

where $D = (u^2 + 4v)^{\frac{1}{2}}$, with corresponding right eigenvectors

$$r_1 = (-1, \frac{u+D}{2})^T, \quad r_2 = (1, \frac{-u+D}{2})^T \qquad (7.0.6)$$

and

$$\begin{cases} \nabla \lambda_1 \cdot r_1 (\frac{3}{2} - \frac{u}{2D}, -\frac{1}{D})(-1, \frac{u+D}{2})^T = -2, \\[4mm] \nabla \lambda_2 \cdot r_2 (\frac{3}{2} + \frac{u}{2D}, \frac{1}{D})(1, \frac{-u+D}{2})^T = 2. \end{cases} \qquad (7.0.7)$$

Therefore if we consider the bounded solution in the upper (u, v)-plane $(v \geq 0)$, it follows from (7.0.5) that $\lambda_1 = \lambda_2$ at point $(0,0)$, at which strict hyperbolicity fails to hold. Moreover by (7.0.7), both characteristic fields are genuinely nonlinear by the definitions given in Chapter 4.

The Riemann invariants of (7.0.1) are functions $w(u, v)$ and $z(u, v)$ satisfying the equations

$$w_u - w_v \frac{u+D}{2} = 0, \quad z_u - z_v \frac{u-D}{2} = 0. \qquad (7.0.8)$$

One solution of (7.0.8) is

$$w(u, v) = u + D, \quad z(u, v) = u - D. \qquad (7.0.9)$$

Consider the Cauchy problem for the related parabolic system

$$\begin{cases} u_t + (u^2 + v)_x = \varepsilon u_{xx}, \\ v_t + (uv)_x = \varepsilon v_{xx} \end{cases} \qquad (7.0.10)$$

with the initial data

$$(u^\varepsilon(x, 0), v^\varepsilon(x, 0)) = (u_0^\varepsilon(x), v_0^\varepsilon(x)), \qquad (7.0.11)$$

where

$$(u_0^\varepsilon(x), v_0^\varepsilon(x)) = (u_0(x), v_0(x) + \varepsilon) * G^\varepsilon \qquad (7.0.12)$$

and G^ε is a mollifier. Then

$$(u_0^\varepsilon(x), v_0^\varepsilon(x)) \in C^\infty \times C^\infty, \qquad (7.0.13)$$

$$(u_0^\varepsilon(x), v_0^\varepsilon(x)) \to (u_0(x), v_0(x)) \text{ a.e., as } \varepsilon \to 0, \qquad (7.0.14)$$

and

$$\|u_0^\varepsilon(x)\| \le M_1, \quad \varepsilon \le v_0^\varepsilon(x) \le M_1, \qquad (7.0.15)$$

for a suitable large constant M_1, which depends only on the L^∞ bound of $(u_0(x), v_0(x))$, but is independent of ε.

We have the main result in this chapter:

Theorem 7.0.1 *Let the initial data $(u_0(x), v_0(x))$ be bounded measurable and $v_0(x) \ge 0$. Then for fixed $\varepsilon > 0$, the viscosity solution $(u^\varepsilon(x,t), v^\varepsilon(x,t))$ of the Cauchy problem (7.0.10), (7.0.11) exists and satisfies*

$$\|u^\varepsilon(x,t)\| \le M_2, \quad 0 < c(\varepsilon, t) \le v^\varepsilon(x,t) \le M_2, \qquad (7.0.16)$$

where M_2 is a positive constant, being independent of ε and $c(\varepsilon, t)$ is a positive function, which could tend to zero as ε tends to zero or t tends to infinity. Moreover, there exists a subsequence (still labelled) $(u^\varepsilon(x,t), v^\varepsilon(x,t))$ such that

$$(u^\varepsilon(x,t), v^\varepsilon(x,t)) \to (u(x,t), v(x,t)), \quad a.e. \text{ on } \Omega, \qquad (7.0.17)$$

where $\Omega \subset R \times R^+$ is any bounded and open set and the pair of limit functions $(u(x,t), v(x,t))$ is a weak solution of the Cauchy problem (7.0.1), (7.0.2).

The first part in Theorem 7.0.1 about the existence of viscosity solutions is proved in Section 7.1; and the second part about the strong convergence (7.0.17) of a subsequence of $(u^\varepsilon(x,t), v^\varepsilon(x,t))$ will be proved in Sections 7.2-7.4.

7.1 Existence of Viscosity Solutions

In this section, we shall give the proof of the existence of viscosity solutions for the Cauchy problem (7.0.10), (7.0.11) and related estimates stated in (7.0.16).

By simple calculations, we have

$$w_u = 1 + \frac{u}{D}, \quad w_v = \frac{2}{D}, \quad w_{uu} = \frac{4v}{D^3}, \quad w_{uv} = -\frac{2u}{D^3}, \quad w_{vv} = -\frac{4}{D^3},$$

and

$$z_u = 1 - \frac{u}{D}, \quad z_v = -\frac{2}{D}, \quad z_{uu} = -\frac{4v}{D^3}, \quad z_{uv} = \frac{2u}{D^3}, \quad z_{vv} = \frac{4}{D^3}.$$

Then multiplying system (7.0.10) by (w_u, w_v) and (z_u, z_v), respectively, we have

$$w(u,v)_t + \lambda_2 w(u,v)_x$$

$$= \varepsilon(w_u u_{xx} + w_v v_{xx})$$

$$= \varepsilon w(u,v)_{xx} - \varepsilon(w_{uu} u_x^2 + 2 w_{uv} u_x v_x + w_{vv} v_x^2)$$

$$= \varepsilon w(u,v)_{xx} - \varepsilon(\frac{4v}{D^3} u_x^2 - \frac{4u}{D^3} u_x v_x - \frac{4}{D^3} v_x^2)$$

$$= \varepsilon w(u,v)_{xx} - \frac{\varepsilon}{D^3}((D+u)u_x + 2v_x)((D-u)u_x - 2v_x)$$

$$= \varepsilon w(u,v)_{xx} - \frac{\varepsilon}{D} w(u,v)_x z(u,v)_x$$

$$\tag{7.1.1}$$

and

$$z(u,v)_t + \lambda_1 z(u,v)_x$$

$$= \varepsilon(z_u u_{xx} + z_v v_{xx})$$

$$= \varepsilon z(u,v)_{xx} - \varepsilon(z_{uu} u_x^2 + 2 z_{uv} u_x v_x + z_{vv} v_x^2)$$

$$= \varepsilon z(u,v)_{xx} + \varepsilon(\frac{4v}{D^3} u_x^2 - \frac{4u}{D^3} u_x v_x - \frac{4}{D^3} v_x^2)$$

$$= \varepsilon z(u,v)_{xx} + \frac{\varepsilon}{D^3}((D+u)u_x + 2v_x)((D-u)u_x - 2v_x)$$

$$= \varepsilon z(u,v)_{xx} + \frac{\varepsilon}{D} w(u,v)_x z(u,v)_x.$$

$$\tag{7.1.2}$$

Now consider (7.1.1) as an equation of the variable w, and (7.1.2) as an equation of the variable z. Applying the maximum principle to (7.1.1) and (7.1.2) respectively, we have that $w(u^\varepsilon, v^\varepsilon) \leq M, z(u^\varepsilon, v^\varepsilon) \geq -M$ if the initial data satisfy $w(u_0^\varepsilon(x), v_0^\varepsilon(x)) \leq M, z(u_0^\varepsilon(x), v_0^\varepsilon(x)) \geq -M$. Thus

$$\Sigma_2 = \{(u,v) : w(u,v) \leq M, z(u,v) \geq -M, v \geq 0\} \tag{7.1.3}$$

is an invariant region for a suitable large constant M (see Figure 7.1).

The positive lower bound estimate about v^ε in (7.0.16) is a direct corollary of the last part of Theorem 1.0.2.

Thus we get the proof of the existence of viscosity solutions of the Cauchy problem (7.0.10), (7.0.11) and the estimates in (7.0.16).

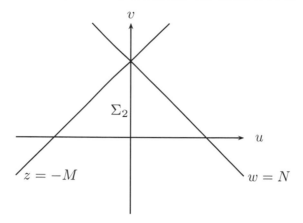

FIGURE 7.1

7.2 Entropy-Entropy Flux Pairs of Lax Type

In this section, four classes of entropy-entropy flux pair of Lax type of the following special forms are constructed:

$$\eta_k^1 = c^{kw}(a_1(D) + \frac{b_1(D,k)}{k}), \quad q_k^1 = e^{kw}(c_1(D) + \frac{d_1(D,k)}{k});$$

$$\eta_{-k}^2 = e^{-kw}(a_2(D) + \frac{b_2(D,k)}{k}), \quad q_{-k}^2 = e^{-kw}(c_2(D) + \frac{d_2(D,k)}{k});$$

$$\eta_{-k}^1 = e^{-kz}(a_3(D) + \frac{b_3(D,k)}{k}), \quad q_{-k}^1 = e^{-kz}(c_3(D) + \frac{d_3(D,k)}{k});$$

$$\eta_k^2 = e^{kz}(a_4(D) + \frac{b_4(D,k)}{k}), \quad q_k^2 = e^{kz}(c_4(D) + \frac{d_4(D,k)}{k}),$$

where w, z are the Riemann invariants of system (7.0.1), and the required estimates on $a_i, b_i (i = 1, 2, 3, 4)$ are obtained by the estimates on the solutions of the Fuchsian equation (6.2.21) given in Lemma 6.2.1.

Let $\rho = D^3, \theta = \frac{3}{2}u$. Then for smooth solutions, (7.0.1) is equivalent to the following system:

$$\begin{cases} \rho_t + (\rho\theta)_x = 0 \\ \theta_t + (\frac{\theta^2}{2} + \frac{3}{8}\rho^{\frac{2}{3}})_x = 0. \end{cases} \tag{7.2.1}$$

Considering the entropy-entropy flux pair (η, q) of system (7.0.1) as functions of variables (ρ, θ), we have

$$(q_\rho, q_\theta) = (\theta\eta_\rho + \frac{1}{4}\rho^{-\frac{1}{3}}\eta_\theta, \rho\eta_\rho + \theta\eta_\theta). \tag{7.2.2}$$

Eliminating the q from (7.2.2), we have

$$\eta_{\rho\rho} = \frac{1}{4}\rho^{-\frac{4}{3}}\eta_{\theta\theta}. \tag{7.2.3}$$

If k denotes a constant, then the function $\eta = h(\rho)e^{k\theta}$ solves (7.2.3) provided that

$$h''(\rho) = \frac{1}{4}k^2\rho^{-\frac{4}{3}}h(\rho). \tag{7.2.4}$$

Let $h(\rho) = \rho^{\frac{1}{3}}\phi(s), s = \frac{3}{2}k\rho^{\frac{1}{3}}$. Then ϕ solves the Fuchsian equation

$$\phi'' - (1 + \frac{2}{s^2})\phi = 0. \tag{7.2.5}$$

Because it is exactly analogous to that in Chapter 6, we may use the method of Frobenius to give a series solution to Equation (7.2.5) as follows:

$$\phi_1 = s^2 \sum_{n=0}^{\infty} c_{2n}s^{2n} = s^2g(s), \tag{7.2.6}$$

where

$$g(s) = \sum_{n=0}^{\infty} c_{2n}s^{2n}, \quad c_{2n} = \frac{c_{2(n-1)}}{(2+2n)(1+2n)+2} \tag{7.2.7}$$

and c_0 is an arbitrary positive constant. Let another independent solution ϕ_2 of (7.2.5) satisfy $\phi_2 = \phi_1 P$. Then P solves

$$P''\phi_1 + 2\phi_1'P' = 0. \tag{7.2.8}$$

Thus

$$P' = -(\phi_1)^{-2} = -(s^4g^2(s))^{-1},$$

and one special function P is given by

$$P = \int_s^{\infty} (s^4g^2(s))^{-1}ds. \tag{7.2.9}$$

Then

$$\phi_2 = s^2 g(s) \int_s^\infty (s^4 g^2(s))^{-1} ds. \tag{7.2.10}$$

It is clear that ϕ_1, ϕ_2 given in (7.2.6) and (7.2.10) satisfy that $\phi_1(s) > 0, \phi_1'(s) > 0$ and $\phi_2(s) > 0$ for all $s \geq 0$. The strict positivity of $\phi_2''(s)$ gives $\phi_2'(s) < 0$ as $s > 0$ because $\lim_{s \to \infty} \phi_2(s) = 0, \lim_{s \to \infty} \phi_2'(s) = 0$.

By simple calculations, the two eigenvalues of (7.2.1) are

$$\lambda_1 = \theta - \frac{\rho^{\frac{1}{3}}}{2}, \quad \lambda_2 = \theta + \frac{\rho^{\frac{1}{3}}}{2} \tag{7.2.11}$$

with corresponding two Riemann invariants

$$z = \theta - \frac{3}{2}\rho^{\frac{1}{3}}, \quad w = \theta + \frac{3}{2}\rho^{\frac{1}{3}}. \tag{7.2.12}$$

Similar to (6.2.5), we have

$$q_w = \lambda_2 \eta_w, \quad q_z = \lambda_1 \eta_z. \tag{7.2.13}$$

From (7.2.12), we have $\theta_w = \frac{1}{2}, \theta_z = \frac{1}{2}, \rho_w = \rho^{\frac{2}{3}}, \rho_z = -\rho^{\frac{2}{3}}$.

Then

$$\begin{cases} q_w = \dfrac{1}{2}q_\theta + \rho^{\frac{2}{3}}q_\rho, \quad q_z = \dfrac{1}{2}q_\theta - \rho^{\frac{2}{3}}q_\rho \\ \eta_w = \dfrac{1}{2}\eta_\theta + \rho^{\frac{2}{3}}\eta_\rho, \quad \eta_z = \dfrac{1}{2}\eta_\theta - \rho^{\frac{2}{3}}\eta_\rho. \end{cases} \tag{7.2.14}$$

From (7.2.14), (7.2.13) and (7.2.11), we have

$$q_\theta = q_w + q_z = \theta\eta_\theta + \rho\eta_\rho. \tag{7.2.15}$$

Letting $\eta_k = \rho^{\frac{1}{3}}\phi(s)e^{k\theta}, \eta_{-k} = \rho^{\frac{1}{3}}\phi(s)e^{-k\theta}$, we have from (7.2.15) that

$$\begin{cases} q_k = \eta_k\left(\theta - \dfrac{2}{3k} + \dfrac{\rho^{\frac{1}{3}}\phi'(s)}{2\phi(s)}\right), \\ q_{-k} = \eta_{-k}\left(\theta + \dfrac{2}{3k} - \dfrac{\rho^{\frac{1}{3}}\phi'(s)}{2\phi(s)}\right). \end{cases} \tag{7.2.16}$$

Let $\eta_k^1 = \rho^{\frac{1}{3}}\phi_1(s)e^{k\theta}$. Then Lemma 6.2.1 gives

$$\eta_k^1 = \rho^{\frac{1}{3}}e^{kw}(1 + O(\frac{1}{s}))\rho^{\frac{1}{3}}e^{kw}(1 + O(\frac{1}{k})) \tag{7.2.17}$$

on any compact subset of $s > 0$.

$$q_k^1 = \eta_k^1(\theta - \frac{2}{3k} + \frac{\rho^{\frac{1}{3}}}{2}(\frac{\phi_1'(s)}{\phi_1(s)} - 1) + \frac{\rho^{\frac{1}{3}}}{2}) = \eta_k^1(\lambda_2 + O(\frac{1}{k})) \quad (7.2.18)$$

on $s \geq 0$ by the fact that $s(\frac{\phi_1'}{\phi_1} - 1)$ is uniformly bounded. Moreover

$$q_k^1 = \eta_k^1(\lambda_2 - \frac{2}{3k} + \frac{s}{3k}(\frac{\phi_1'(s)}{\phi_1(s)} - 1)) = \eta_k^1(\lambda_2 - \frac{2}{3k} + O(\frac{1}{k^2})) \quad (7.2.19)$$

on any compact subset of $s > 0$. Similarly

$$\eta_k^2 = \rho^{\frac{1}{3}}\phi_2(s)e^{k\theta}\rho^{\frac{1}{3}}e^{kz}(1 + O(\frac{1}{k})) \quad (7.2.20)$$

on any compact subset of $s > 0$,

$$q_k^2 = \eta_k^2(\lambda_1 + O(\frac{1}{k})) \quad (7.2.21)$$

on $s \geq 0$ and

$$q_k^2 = \eta_k^2(\lambda_1 - \frac{2}{3k} + O(\frac{1}{k^2})) \quad (7.2.22)$$

on any compact subset of $s > 0$.

$$\eta_{-k}^1 = \rho^{\frac{1}{3}}\phi_1(s)e^{-k\theta}\rho^{\frac{1}{3}}e^{-kz}(1 + O(\frac{1}{k})) \quad (7.2.23)$$

on any compact subset of $s > 0$,

$$q_{-k}^1 = \eta_{-k}^1(\lambda_1 + O(\frac{1}{k})) \quad (7.2.24)$$

on $s \geq 0$ and

$$q_{-k}^1 = \eta_{-k}^1(\lambda_1 + \frac{2}{3k} + O(\frac{1}{k^2})) \quad (7.2.25)$$

on any compact subset of $s > 0$. Finally

$$\eta_{-k}^2 = \rho^{\frac{1}{3}}\phi_2(s)e^{-k\theta}\rho^{\frac{1}{3}}e^{-kw}(1 + O(\frac{1}{k})) \quad (7.2.26)$$

on any compact subset of $s > 0$,

$$q_{-k}^2 = \eta_{-k}^2(\lambda_2 + O(\frac{1}{k})) \quad (7.2.27)$$

on $s \geq 0$ and

$$q_{-k}^2 = \eta_{-k}^2(\lambda_2 + \frac{2}{3k} + O(\frac{1}{k^2})) \qquad (7.2.28)$$

on any compact subset of $s > 0$.

The estimates (7.2.17)-(7.2.28) will be used in Section 7.4 to reduce the Young measures, corresponding to the sequence of viscosity solutions, to be Dirac measures, and hence the proof of the existence of weak solutions in the second part of Theorem 7.0.1.

7.3 Compactness of $\eta_t + q_x$ in H_{loc}^{-1}

In this section, we mainly prove the following theorem:

Theorem 7.3.1 *For the entropy-entropy flux pairs* (η, q) *of Lax type constructed in Section 7.2,*

$$\eta(u^\varepsilon, v^\varepsilon)_t + q(u^\varepsilon, v^\varepsilon)_x$$

is compact in $H_{loc}^{-1}(R \times R^+)$ *with respect to the viscosity solutions* $(u^\varepsilon, v^\varepsilon)$ *obtained in Section 7.1.*

Proof. It is obvious that system (7.0.1) has a convex entropy $\eta^\star(u, v) = \frac{u^2}{2} + \int_0^v \log v dv$ and the corresponding entropy flux $q^\star(u, v) = \frac{2u^3}{3} + uv \log v$. Similar to the proof of (6.3.2), we have

$$\varepsilon^{\frac{1}{2}}\partial_x u^\varepsilon \text{ and } \varepsilon^{\frac{1}{2}}\frac{1}{v^\varepsilon}\partial_x v^\varepsilon \text{ are uniformly bounded in } L_{loc}^2(R \times R^+).$$
$$(7.3.1)$$

We only prove Theorem 7.3.1 for the entropy-entropy flux pairs (η_k^2, q_k^2). A similar method may give the proofs for other entropy-entropy flux pairs (η_k^1, q_k^1), (η_{-k}^1, q_{-k}^1) and (η_{-k}^2, q_{-k}^2).

Multiplying system (7.0.10) by (η_u, η_v), we have

$$\eta_t + q_x = \varepsilon\eta_{xx} - \varepsilon(\eta_{uu}u_x^2 + 2\eta_{uv}u_x v_x + \eta_{vv}v_x^2). \qquad (7.3.2)$$

Because

$$\int_s^\infty \frac{ds}{s^4 g^2(s)} = O(\frac{1}{s^3}) \qquad (7.3.3)$$

as s approaches zero, for any fixed $k > 0$ we have that

$$\eta_k^2 = \frac{2}{3k}s^3g(s)\int_s^\infty \frac{ds}{s^4g^2(s)}e^{k\theta} \tag{7.3.4}$$

and q_k^2 both are uniformly bounded. Thus $(\eta_k^2(u,v))_t + (q_k^2(u,v))_x$ is bounded in $W^{-1,\infty}(R \times R^+)$.

Because

$$\eta_k^2 = \frac{2e^{k\theta}}{k}\left(\frac{1}{g(s)} - 2s^3g(s)\int_s^\infty \frac{g'(s)ds}{s^3g^3(s)}\right) \tag{7.3.5}$$

and $g'(s)/s \le g(s)$, we have

$$\int_s^\infty \frac{g'(s)ds}{s^3g^3(s)} = O(\frac{1}{s}) \tag{7.3.6}$$

as s approaches zero. Thus $(\eta_k^2)_s$ and $(\eta_k^2)_\theta$ both are bounded and $(\eta_k^2)_s = O(s)$ as s approaches zero. Since

$$\frac{\partial s}{\partial u} = 3k(v^2 + 4u)^{-\frac{1}{2}}, \quad \frac{\partial s}{\partial \theta} = \frac{3kv}{2}(v^2 + 4u)^{-\frac{1}{2}}, \tag{7.3.7}$$

then $\varepsilon\partial_x\eta_k^2 = O(\varepsilon(|u_x|+|v_x|))$. Thus $\varepsilon\partial_{xx}\eta_k^2$ is compact in $H_{loc}^{-1}(R\times R^+)$ from (7.3.1).

The proof of Theorem 7.3.1 will be completed if we can prove that the second term in the right-hand side of (7.3.2) is bounded in $L_{loc}^1(R \times R^+)$.

Since $\eta_k^2 = I_1 - \frac{4}{k}e^{k\theta}I$, where $I_1 = \frac{2}{kg(s)}e^{k\theta}$ is uniformly bounded in C^2, and $I = s^3g(s)\int_s^\infty \frac{g'(s)ds}{s^3g^3}$, then the proof of Theorem 7.3.1 is concentrated to the boundedness of L in $L_{loc}^1(R \times R^+)$, where

$$L = \varepsilon(I_{uu}u_x^2 + 2I_{uv}u_xv_x + I_{vv}v_x^2). \tag{7.3.8}$$

Let $L = L_1 + L_2$, where

$$\begin{cases} L_1 = \varepsilon I_{ss}((s_u)^2u_x^2 + 2s_us_vu_xv_x + (s_v)^2v_x^2) \\ L_2 = \varepsilon I_s(s_{uu}u_x^2 + 2s_{uv}u_xv_x + s_{vv}v_x^2). \end{cases} \tag{7.3.9}$$

Because $g'(s)/s \le g(s), I_s = O(s)$ as s approaches zero and I_{ss} is bounded, we have that L_1 and L_2 are controlled by $\varepsilon(O(\frac{v_x^2}{|v|}) + O(u_x^2))$ and hence bounded in $L_{loc}^1(R \times R^+)$ from (7.3.1). This completes the proof of Theorem 7.3.1. ∎

7.4 Existence of Weak Solutions

In this section, we shall use the compensated compactness method to prove the existence of weak solutions of the Cauchy problem (7.0.1), (7.0.10) given in the second part of Theorem 7.0.1.

Consider a compactly supported probability measure ν on R^2 such that

$$< \nu, \eta^1 >< \nu, q^2 > - < \nu, \eta^2 >< \nu, q^1 >< \nu, \eta^1 q^2 - \eta^2 q^1 > \quad (7.4.1)$$

for entropy-entropy flux pairs $(\eta^i, q^i)(i = 1, 2)$, of system (7.0.1), satisfying the compactness of $\eta^i(u^\varepsilon, v^\varepsilon)_t + q^i(u^\varepsilon, v^\varepsilon)_x$ in $H^{-1}(\Omega)$. Then the proof of the existence of weak solutions in Theorem 7.0.1 is reduced to proving that ν is a point mass. We shall realize this goal by the entropy-entropy flux pairs $(\eta^i_{\pm k}, q^i_{\pm k})$ constructed in Section 7.2.

The proof is almost the same as that given in Section 6.4.

Let Q denote the smallest characteristic rectangle

$$Q = \{(u, v) : w^- \leq w \leq w^+, \quad z^- \leq z \leq z^+, \quad v \geq 0\}.$$

Then w^-, w^+ are nonnegative and z^-, z^+ are nonpositive.

If the support of ν only consists exactly of the point $(0, 0)$, we are done. Next, consider the other case, where the support of ν is not concentrated at the point $(u, v) = (0, 0)$, so that $< \nu, \eta^1_k >> 0$ and $< \nu, \eta^2_{-k} >> 0$.

We introduce probability measures $\mu^\pm_k$ on Q defined by

$$< \mu^+_k, h >=< \nu, h\eta^1_k > / < \nu, \eta^1_k > \quad (7.4.2)$$

and

$$< \mu^-_k, h >=< \nu, h\eta^1_{-k} > / < \nu, \eta^1_{-k} >, \quad (7.4.3)$$

where $h = h(u, v)$ denotes an arbitrary continuous function. As a consequence of weak-star compactness, there exist probability measures $\mu^\pm$ on Q such that

$$< \mu^\pm, h >= \lim_{k \to \infty} < \mu^\pm_k, h > \quad (7.4.4)$$

after the selection of an appropriate subsequence. We observe that the measures μ^+, μ^- are respectively concentrated on the boundary sections of Q associated with w, i.e.,

$$Q \bigcap \{(u, v) : w = w^+\} \text{ and } Q \bigcap \{(u, v) : w = w^-\}. \tag{7.4.5}$$

Because

$$\eta_k^1 = e^{kw} \rho^{\frac{1}{3}} (1 + O(\frac{1}{k})), \quad \text{and } \eta_{-k}^2 = e^{-kw} \rho^{\frac{1}{3}} (1 + O(\frac{1}{k})), \tag{7.4.6}$$

on any compact subset of $s > 0$ and by the assumption, the support of ν is not concentrated at $s = 0$.

Similar to the proof of (6.4.15), we have that

$$< \mu^+, \lambda_2 \eta - q > = < \mu^-, \lambda_2 \eta - q > \tag{7.4.7}$$

for any (η, q) satisfying that $\eta_t + q_x$ is compact in $H_{loc}^{-1}(R \times R^+)$.

Substituting (η_k^1, q_k^1) into (7.4.7) yields that

$$< \mu^-, \lambda_2 \eta_k^1 - q_k^1 > \le c_1 e^{kw^-} / k \tag{7.4.8}$$

and

$$< \mu^+, \lambda_2 \eta_k^1 - q_k^1 > \ge c_2 e^{kw^+} / k, \tag{7.4.9}$$

where c_1, c_2 are two positive constants. Therefore (7.4.7), (7.4.8) and (7.4.9) imply that $w^- = w^+$.

In the same fashion we conclude that $z^- = z^+$. This completes the proof of Theorem 7.0.1. ∎

7.5   Related Results

Two characteristic fields of system (7.0.1) or the curves determined by the equations $w = const, z = const$ in the (u, v)-plane are clearly straight lines, so system (7.0.1) is of Temple type [Te], whose shock curves and rarefaction curves coincided. The global existence and uniqueness of an L^∞ weak solution for a general $n \times n$ system of Temple type in strictly hyperbolic regions were established by Heibig in [He].

In nonstrictly hyperbolic regions, a compact framework for an $n \times n$ system of chromatography

$$u_{it} + \left(\frac{k_i u_i}{1 + D}\right)_x = 0, \quad i = 1, 2, \cdots, n, \tag{7.5.1}$$

was established by James, Peng and Perthame [JPP] by using the kinetic formulation coupled with the compensated compactness method, where k_i are positive constants satisfying $0 < k_1 < k_2 < \cdots < k_n$ and $D = 1 + k_1 + k_2 + \cdots + k_n$.

System (7.5.1) is the unique application of the compensated compactness method on hyperbolic systems of more than two equations. However, it should be a very interesting topic to construct suitable approximated solutions $\{u_i^l\}$ of system (7.5.1) and then to prove the compactness of $\eta(u_i^l)_t + q(u_i^l)_x$ in $W_{loc}^{-1,2}$, for the entropy-entropy flux pairs (η, q) constructed by the kinetic formulation in [JPP], with respect to the sequence $\{u_i^l\}$. If this is done, then the existence of weak solutions to system (7.5.1) follows from the compactness framework given in [JPP]. The proof of Theorem 7.0.1 is from the paper [LMR]. The main difficulty in dealing with system (7.0.1) is the singularity of entropies at the nonstrictly hyperbolic domain. System (7.5.1) should be more difficult since it has more equations.

Chapter 8

System of Polytropic Gas Dynamics

We consider the Cauchy problem for the system of isentropic gas dynamics in Eulerian coordinates

$$\begin{cases} \rho_t + (\rho u)_x = 0 \\ (\rho u)_t + (\rho u^2 + P(\rho))_x = 0, \end{cases} \qquad (8.0.1)$$

with bounded measurable initial data

$$(\rho(x,0), u(x,0)) = (\rho_0(x), u_0(x)), \quad \rho_0(x) \geq 0, \qquad (8.0.2)$$

where ρ is the density of gas, u the velocity, $P = P(\rho)$ the pressure satisfying $P'(\rho) \geq 0$. For the polytropic gas, P takes the special form $P(\rho) = c\rho^\gamma$, where $\gamma > 1$ and c is an arbitrary positive constant, for instance, $c = k^2 = \frac{(\gamma-1)^2}{4\gamma}$.

System (8.0.1) can be written into

$$\begin{cases} \rho_t + m_x = 0 \\ m_t + (\frac{m^2}{\rho} + P(\rho))_x = 0, \end{cases} \qquad (8.0.3)$$

where m denotes the mass.

Let F be the mapping from R^2 into R^2 defined by

$$F : (\rho, m) \rightarrow (m, \frac{m^2}{\rho} + P(\rho)).$$

Then

$$dF = \begin{pmatrix} 0 & 1 \\ -\frac{m^2}{\rho} + P'(\rho) & \frac{2m}{\rho} \end{pmatrix}, \tag{8.0.4}$$

and the eigenvalues of system (8.0.1) are solutions of the following characteristic equation:

$$\lambda^2 - \frac{2m}{\rho}\lambda + \frac{m^2}{\rho} - P'(\rho) = 0. \tag{8.0.5}$$

Thus two eigenvalues of system (8.0.1) are

$$\lambda_1 = \frac{m}{\rho} - \sqrt{P'(\rho)}, \quad \lambda_2 = \frac{m}{\rho} + \sqrt{P'(\rho)} \tag{8.0.6}$$

with corresponding right eigenvectors

$$r_1 = (1, \lambda_1)^T, \quad r_2 = (1, \lambda_2)^T. \tag{8.0.7}$$

The Riemann invariants of (8.0.1) are functions $w(\rho, m)$ and $z(\rho, m)$ satisfying the equations

$$(w_\rho, w_m) \cdot dF = \lambda_2(w_\rho, w_m) \quad \text{and} \quad (z_\rho, z_m) \cdot dF = \lambda_1(z_\rho, z_m). \tag{8.0.8}$$

One solution of (8.0.8) is

$$w(u, v) = \frac{m}{\rho} + \int_0^\rho \frac{\sqrt{P'(s)}}{s}ds, \quad z(u, v) = \frac{m}{\rho} - \int_0^\rho \frac{\sqrt{P'(s)}}{s}ds. \tag{8.0.9}$$

By simple calculations,

$$\begin{aligned} \nabla\lambda_1 \cdot r_1 &= (-\frac{m}{\rho^2} - \frac{P''(\rho)}{2\sqrt{P'(\rho)}}, \frac{1}{\rho})(1, \lambda_1)^T \\ &= -\frac{\rho P''(\rho) + 2P'(\rho)}{2\rho\sqrt{P'(\rho)}}, \end{aligned} \tag{8.0.10}$$

and

$$\begin{aligned} \nabla\lambda_2 \cdot r_2 &= (-\frac{m}{\rho^2} + \frac{P''(\rho)}{2\sqrt{P'(\rho)}}, \frac{1}{\rho})(1, \lambda_2)^T \\ &= \frac{\rho P''(\rho) + 2P'(\rho)}{2\rho\sqrt{P'(\rho)}}. \end{aligned} \tag{8.0.11}$$

For the case of polytropic gas, we get

$$\lambda_1 = \frac{m}{\rho} - \theta \rho^\theta, \quad \lambda_2 = \frac{m}{\rho} + \theta \rho^\theta, \tag{8.0.12}$$

$$w(u,v) = \frac{m}{\rho} + \rho^\theta, \quad z(u,v) = \frac{m}{\rho} - \rho^\theta, \tag{8.0.13}$$

where $\theta = \frac{\gamma-1}{2}$, and

$$\nabla \lambda_1 \cdot r_1 = -\frac{(\gamma-1)(\gamma+1)}{4}\rho^{\gamma-3}, \quad \nabla \lambda_2 \cdot r_2 = \frac{(\gamma-1)(\gamma+1)}{4}\rho^{\gamma-3}. \tag{8.0.14}$$

Therefore it follows from (8.0.12) that for the case of polytropic gas, system (8.0.1) is strictly hyperbolic in the domain $\{(x,t) : \rho(x,t) > 0\}$, while it is nonstrictly hyperbolic in the domain $\{(x,t) : \rho(x,t) = 0\}$, since $\lambda_1 = \lambda_2$ when $\rho = 0$. From (8.0.14), system (8.0.1) is genuinely nonlinear if the adiabatic exponent $\gamma \in (1,3]$, while the system is no longer genuinely nonlinear at $\rho = 0$ if the adiabatic exponent $\gamma > 3$.

Consider the Cauchy problem for the related parabolic system

$$\begin{cases} \rho_t + m_x = \varepsilon \rho_{xx} \\ m_t + (\frac{m^2}{\rho} + P(\rho))_x = \varepsilon m_{xx}, \end{cases} \tag{8.0.15}$$

with initial data

$$(\rho^\varepsilon(x,0), m^\varepsilon(x,0)) = (\rho_0^\varepsilon(x), m_0^\varepsilon(x)), \tag{8.0.16}$$

where

$$(\rho_0^\varepsilon(x), m_0^\varepsilon(x)) = (\rho_0(x) + \varepsilon, \rho_0(x)u_0(x)) * G^\varepsilon \tag{8.0.17}$$

and G^ε is a mollifier. Then

$$(\rho_0^\varepsilon(x), m_0^\varepsilon(x)) \in C^\infty \times C^\infty, \tag{8.0.18}$$

$$(\rho_0^\varepsilon(x), m_0^\varepsilon(x)) \to (\rho_0(x), m_0(x)) \text{ a.e., as } \varepsilon \to 0, \tag{8.0.19}$$

and

$$\varepsilon \le \rho_0^\varepsilon(x) \le M_1, \quad \|u_0^\varepsilon(x)\| = \|\frac{m_0^\varepsilon(x)}{\rho_0^\varepsilon(x)}\| \le M_1, \tag{8.0.20}$$

for a suitable large constant M_1, which depends only on the L^∞ bound of $(\rho_0(x), u_0(x))$, but is independent of ε.

We have the main result in this chapter as follows:

Theorem 8.0.1 *Let the initial data* $(\rho_0(x), u_0(x))$ *be bounded measurable and* $\rho_0(x) \geq 0, P(\rho) \in C^2(0, \infty), P'(\rho) > 0, 2P'(\rho) + \rho P''(\rho) \geq 0$ *for* $\rho > 0;$ *and*

$$\int_c^{\infty} \frac{\sqrt{P'(\rho)}}{\rho} d\rho = \infty, \quad \int_0^c \frac{\sqrt{P'(\rho)}}{\rho} d\rho < \infty, \quad \forall c > 0.$$

Then for fixed $\varepsilon > 0$, *the smooth viscosity solution* $(\rho^\varepsilon(x, t), m^\varepsilon(x, t))$ *of the Cauchy problem* (8.0.15), (8.0.16) *exists and satisfies*

$$0 < c(\varepsilon, t) \leq \rho^\varepsilon(x, t) \leq M_2, \quad \|u^\varepsilon(x, t)\| = \|\frac{m^\varepsilon(x, t)}{\rho^\varepsilon(x, t)}\| \leq M_2, \quad (8.0.21)$$

where M_2 *is a positive constant, being independent of* ε; $c(\varepsilon, t)$ *is a positive function, which could tend to zero as* ε *tends to zero or* t *tends to infinity.*

Moreover, for the polytropic gas and $\gamma > 1$, *there exists a subsequence (still labelled)* $(\rho^\varepsilon(x, t), \rho^\varepsilon(x, t)u^\varepsilon(x, t))$ *which converges almost everywhere on any bounded and open set* $\Omega \subset R \times R^+$:

$$(\rho^\varepsilon(x, t), \rho^\varepsilon(x, t)u^\varepsilon(x, t)) \to (\rho(x, t), \rho(x, t)u(x, t)), \quad as \ \varepsilon \downarrow 0^+,$$
$$(8.0.22)$$

where the limit pair of functions $(\rho(x, t), \rho(x, t)u(x, t))$ *is a weak solution of the Cauchy problem* (8.0.1), (8.0.2).

The existence of viscosity solutions and related estimates (8.0.21) shall be proved in Section 8.1. For the case of adiabatic exponent $\gamma > 3$, the strong convergence (8.0.22) of a subsequence of $(\rho^\varepsilon(x, t), m^\varepsilon(x, t))$ is proved in Sections 8.3-8.4. Finally, the proof of the strong convergence subsequence of $(\rho^\varepsilon(x, t), m^\varepsilon(x, t))$ for the case of $1 < \gamma \leq 3$ is given in Section 8.5, where we introduce a different short proof for this case, but with two more assumptions (A_1) and (A_2) on viscosity solutions:

(A_1) The initial data $(\rho_0(x), u_0(x))$ is small;

(A_2) The viscosity solutions $(\rho^\varepsilon(x, t), m^\varepsilon(x, t))$ of the Cauchy problem (8.0.15)-(8.0.16) have an *a priori* estimate $\rho^\varepsilon(x, t) \geq c(t) > 0$, where $c(t)$ is independent of ε, but could tend to zero as t tends to infinity.

Remark 8.0.2 *The condition* (A_1) *ensures that the viscosity solutions are small and so the support sets of the Young measures introduced in Theorem 2.2.1 are also small. The condition* (A_2) *is remarked in [LPS], p. 629.*

8.1   Existence of Viscosity Solutions

To prove the existence of smooth viscosity solutions $(\rho^\varepsilon(x,t), m^\varepsilon(x,t))$ for the Cauchy problem (8.0.15), (8.0.16), by Theorem 1.0.2, we only need to prove the *a priori* estimates given in (8.0.21).

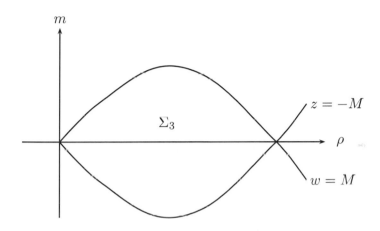

FIGURE 8.1

We multiply (8.0.15) by (w_ρ, w_m) and (z_ρ, z_m), respectively, to obtain

$$w_t + \lambda_2 w_x$$

$$= \varepsilon w_{xx} + \frac{2\varepsilon}{\rho}\rho_x w_x - \frac{\varepsilon}{2\rho^2\sqrt{P'(\rho)}}(2P' + \rho P'')\rho_x^2, \tag{8.1.1}$$

and

$$z_t + \lambda_1 z_x$$

$$= \varepsilon z_{xx} + \frac{2\varepsilon}{\rho}\rho_x z_x + \frac{\varepsilon}{2\rho^2\sqrt{P'(\rho)}}(2P' + \rho P'')\rho_x^2. \tag{8.1.2}$$

Then the assumptions on $P(\rho)$ yield

$$w_t + \lambda_2 w_x \leq \varepsilon w_{xx} + \frac{2\varepsilon}{\rho}\rho_x w_x \tag{8.1.3}$$

and

$$z_t + \lambda_1 z_x \geq \varepsilon z_{xx} + \frac{2\varepsilon}{\rho} \rho_x z_x. \tag{8.1.4}$$

If we consider (8.1.3) and (8.1.4) as inequalities about the variables w and z, then we can get the estimates $w(\rho^\varepsilon, m^\varepsilon) \leq M, z(\rho^\varepsilon, m^\varepsilon) \geq -M$ by applying the maximum principle to (8.1.3) and (8.1.4). This shows that the region (see Figure 8.1)

$$\Sigma_3 = \{(\rho, m) : w(\rho, m) \leq M, \ z(\rho, m) \geq -M\}$$

is an invariant region. Thus we obtain the estimates $0 \leq \rho^\varepsilon \leq M_2$ and $\|u^\varepsilon\| \leq M_2$ for a suitable constant M_2, since $\int_c^\infty \frac{\sqrt{P'(\rho)}}{\rho} d\rho = \infty$ and $\int_0^c \frac{\sqrt{P'(\rho)}}{\rho} d\rho < \infty$ for any constant $c > 0$.

Since u is uniformly bounded, the last part of Theorem 1.0.2 gives the positive lower bound of ρ.

Therefore we get the proof of the existence of smooth viscosity solutions for the Cauchy problem (8.0.15) and (8.0.16).

8.2 Weak Entropies and H_{loc}^{-1} Compactness

In this section, we shall first construct the weak entropy-entropy flux pairs (η, q) of system (8.0.1) for the polytropic case $P(\rho) = c\rho^\gamma$ with the exponent $\gamma > 1$ and $c = k^2 = \frac{(\gamma-1)^2}{4\gamma}$, and then prove the compactness of $\eta(\rho^\varepsilon, m^\varepsilon)_t + q(\rho^\varepsilon, m^\varepsilon)_x$ in $H_{loc}^{-1}(R \times R^+)$, with respect to the viscosity approximated solutions $(\rho^\varepsilon(x,t), m^\varepsilon(x,t))$ of the Cauchy problem (8.0.15) and (8.0.16).

Rewriting the second equation in (8.0.1) as

$$\rho_t u + \rho u_t + (\rho u)_x u + \rho u u_x + P(\rho)_x = 0 \tag{8.2.1}$$

and substituting the first equation in (8.0.1) into (8.2.1), we get the following new system:

$$\begin{cases} \rho_t + (\rho u)_x = 0 \\ u_t + (\frac{1}{2}u^2 + \int_0^\rho \frac{P'(s)}{s} ds)_x = 0. \end{cases} \tag{8.2.2}$$

If solutions have shock-waves, system (8.0.1) and system (8.2.2) are different. In Chapter 9 and Chapter 10, we shall study the weak solutions for the Cauchy problem (8.2.2) with bounded measurable initial data.

However, for smooth solutions, system (8.2.2) is equivalent to system (8.0.1), and particularly, both systems have the same entropy-entropy flux pairs. Thus any entropy-entropy flux pair $(\eta(\rho, m), q(\rho, m))$ of system (8.0.1) satisfies the additional system

$$(q_\rho, q_u) = (\eta_\rho, \eta_u) \cdot \begin{pmatrix} u & \rho \\ \frac{P'(\rho)}{\rho} & u \end{pmatrix}, \tag{8.2.3}$$

or equivalently

$$q_\rho = u\eta_\rho + \frac{P'(\rho)}{\rho}\eta_u, \quad q_u = \rho\eta_\rho + u\eta_u. \tag{8.2.4}$$

Eliminating the q from (8.2.4), we have

$$\eta_{\rho\rho} = \frac{P'(\rho)}{\rho^2}\eta_{uu}. \tag{8.2.5}$$

An entropy $\eta(\rho, u)$ of system (8.0.1) is called a weak entropy if $\eta(0, u) = 0$, that is, a solution of Equation (8.2.5) with the special initial conditions:

$$\eta(\rho = 0, u) = 0, \quad \eta_\rho(\rho = 0, u) = g(u), \tag{8.2.6}$$

where $g(u)$ is an arbitrary given function. The solution of (8.2.5)-(8.2.6) is given by the following lemma:

Lemma 8.2.1 *For $\rho \geq 0, u, w \in R$, let*

$$G(\rho, w) = (\rho^{\gamma-1} - w^2)_+^\lambda, \quad \lambda = \frac{3-\gamma}{2(\gamma-1)}, \tag{8.2.7}$$

where the notation $x_+ = \sup(0, x)$. Then the solution of (8.2.5)-(8.2.6) is given by the formula

$$\eta(w, z) = \int_z^w [(w-s)(s-z)]^\lambda g(s)ds$$

$$= (w-z)^{\frac{2}{\gamma-1}} \int_0^1 [\tau(1-\tau)]^\lambda g(w - (w-z)\tau)d\tau \tag{8.2.8}$$

or

$$\eta(\rho, u) = \int_R g(\xi)G(\rho, \xi - u)d\xi$$

$$= 2^{\frac{2}{\gamma-1}}\rho \int_0^1 [\tau(1-\tau)]^\lambda g(u + \rho^\theta - 2\rho^\theta \tau)d\tau; \tag{8.2.9}$$

and the weak entropy flux q of system (8.0.1) associated with η is

$$q(\rho, u) = \int_R g(\xi)[\theta\xi + (1 - \theta)u]G(\rho, \xi - u)d\xi, \tag{8.2.10}$$

where $\theta = \frac{\gamma-1}{2}$. Moreover, if the initial data $g(u) \in C^1(R)$, then the weak entropy η satisfies the following estimates:

$$|\eta_\rho(\rho, m)| \leq M, \quad |\eta_m(\rho, m)| \leq M, \tag{8.2.11}$$

where M depends only on the L^∞ bound M_2 of (ρ, u).

Proof. Exactly analogous to (6.2.6), the entropy of system (8.0.1) also satisfies the following equation:

$$\eta_{wz} + \frac{\lambda_{2z}}{\lambda_2 - \lambda_1}\eta_w - \frac{\lambda_{1w}}{\lambda_2 - \lambda_1}\eta_z = 0, \tag{8.2.12}$$

where $\lambda_1, \lambda_2, w, z$ are given by (8.0.12)-(8.0.13).

By simple calculations, we have

$$\lambda_1 = \frac{3-\gamma}{4}w + \frac{\gamma+1}{4}z, \quad \lambda_2 = \frac{\gamma+1}{4}w + \frac{3-\gamma}{4}z, \tag{8.2.13}$$

$$\lambda_{2z} = \lambda_{1w} = \frac{3-\gamma}{4}, \quad \lambda_2 - \lambda_1 = \frac{\gamma-1}{2}(w - z), \tag{8.2.14}$$

and

$$\eta_\rho(\rho, u) = \theta\rho^{\theta-1}(\eta(w, z)_w - \eta(w, z)_z)$$

$$= \theta(\frac{w-z}{2})^b(\eta(w, z)_w - \eta(w, z)_z), \tag{8.2.15}$$

where $b = \frac{\gamma-3}{\gamma-1}$. Therefore the entropy equation (8.2.12) is reduced to the following Darboux-Euler-Poisson equation:

$$\eta_{wz} + \frac{\lambda}{w-z}(\eta_w - \eta_z) = 0. \tag{8.2.16}$$

The weak entropy η satisfies the conditions

$$\lim_{(w-z)\to 0} \eta(w,z) = 0, \tag{8.2.17}$$

and

$$\lim_{(w-z)\to 0} \theta(\frac{w-z}{2})^{\frac{\gamma-3}{\gamma-1}} (\eta(w,z)_w - \eta(w,z)_z) = g(w). \tag{8.2.18}$$

Therefore, by the theory of Darboux-Euler-Poisson equation (cf. [Bi]), we get that the weak entropy is of the representation formula (8.2.8), which is in the form (8.2.9) if we consider it to be the function of (ρ, u).

Using the second equation in (8.2.4) and the weak solution formula (8.2.9), we have

$$q_u = \eta + 2^{\frac{2}{\gamma-1}}\theta \int_0^1 [\tau(1-\tau)]^{\lambda} g'(u + \rho^{\theta} - 2\rho^{\theta}\tau)(1-2\tau)\rho^{\theta+1} d\tau$$

$$+2^{\frac{2}{\gamma-1}} u \int_0^1 [\tau(1-\tau)]^{\lambda} g'(u + \rho^{\theta} - 2\rho^{\theta}\tau) d\tau. \tag{8.2.19}$$

Since

$$\int^u u g'(u + \rho^{\theta} - 2\rho^{\theta}\tau) du$$

$$ug(u + \rho^{\theta} - 2\rho^{\theta}\tau) - \int^u g(u + \rho^{\theta} - 2\rho^{\theta}\tau) du,$$

we get from (8.2.19) that

$$q = \int^u \eta du + \eta - \int^u \eta du$$

$$+2^{\frac{2}{\gamma-1}}\theta \int_0^1 [\tau(1-\tau)]^{\lambda} g(u + \rho^{\theta} - 2\rho^{\theta}\tau)(1-2\tau)\rho^{\theta+1} d\tau$$

$$= \int_z^w [(w-s)(s-z)]^{\lambda}(1-\theta)u + \theta s)g(s) ds, \tag{8.2.20}$$

and hence the proof of (8.2.10).

Noticing the last part in (8.2.9), we get the boundedness of $\nabla \eta$ and hence, the proof of Lemma 8.2.1. ∎

Lemma 8.2.2 *For any weak entropy* $\eta(\rho, m)$ *of system* (8.0.1),

$$\varepsilon(\rho_x^\varepsilon, m_x^\varepsilon) \cdot \nabla^2 \eta(\rho^\varepsilon, m^\varepsilon) \cdot (\rho_x^\varepsilon, m_x^\varepsilon)^T \tag{8.2.21}$$

is bounded in $L^1_{loc}(R \times R^+)$, *where*

$$\nabla^2 \eta(\rho^\varepsilon, m^\varepsilon) = \begin{pmatrix} \eta_{\rho\rho}(\rho^\varepsilon, m^\varepsilon) & \eta_{\rho m}(\rho^\varepsilon, m^\varepsilon) \\ \eta_{m\rho}(\rho^\varepsilon, m^\varepsilon) & \eta_{mm}(\rho^\varepsilon, m^\varepsilon) \end{pmatrix}. \tag{8.2.22}$$

Proof. For simplicity, we omit the superscript ε in the viscosity solutions $(\rho^\varepsilon, m^\varepsilon)$.

It is easy to check that system (8.0.1) has a convex entropy

$$\eta^\star = \frac{m^2}{2\rho} + \frac{k}{\gamma - 1}\rho^\gamma, \tag{8.2.23}$$

and hence the boundedness of

$$\varepsilon(\rho_x, m_x) \cdot \nabla^2 \eta^\star(\rho, m) \cdot (\rho_x, m_x)^T \tag{8.2.24}$$

in $L^1_{loc}(R \times R^+)$ can be obtained by using the same technique as given in (6.3.2) or (7.3.1). Then it follows that

$$\varepsilon k \gamma \rho^{\gamma-2}(\rho_x)^2 + \varepsilon \frac{1}{\rho}[\frac{m}{\rho}\rho_x - m_x]^2 \tag{8.2.25}$$

is bounded in $L^1_{loc}(R \times R^+)$.

Since

$$k\gamma\rho^{\gamma-2}(\rho_x)^2 + \frac{1}{\rho}[\frac{m}{\rho}\rho_x - m_x]^2 = k\gamma\rho^{\gamma-2}(\rho_x)^2 + \rho(u_x)^2, \tag{8.2.26}$$

we get the boundedness of

$$\varepsilon\rho^{\gamma-2}(\rho_x)^2, \quad \varepsilon\frac{1}{\rho}[\frac{m}{\rho}\rho_x - m_x]^2, \quad \varepsilon\rho(u_x)^2 \tag{8.2.27}$$

in $L^1_{loc}(R \times R^+)$.

By simple calculations, for a weak entropy η represented by (8.2.9), we have that

$$\eta_\rho(\rho, m) = \int_0^1 [\tau(1 - \tau)]^\lambda g(\frac{m}{\rho} + \rho^\theta - 2\rho^\theta\tau)d\tau$$

$$+ \int_0^1 [\tau(1 - \tau)]^\lambda g'(\frac{m}{\rho} + \rho^\theta - 2\rho^\theta\tau)(-\frac{m}{\rho} + (1 - 2\tau)\theta\rho^\theta)d\tau, \tag{8.2.28}$$

$$\eta_m(\rho, m) = \int_0^1 [\tau(1-\tau)]^\lambda g'(\frac{m}{\rho} + \rho^\theta - 2\rho^\theta \tau) d\tau, \qquad (8.2.29)$$

$$\eta_{\rho\rho}(\rho, m)$$

$$= \int_0^1 [\tau(1-\tau)]^\lambda g'(\frac{m}{\rho} + \rho^\theta - 2\rho^\theta \tau)[(1-2\tau)(\theta + \theta^2)\rho^{\theta-1}] d\tau$$

$$+ \int_0^1 [\tau(1-\tau)]^\lambda g''(\frac{m}{\rho} + \rho^\theta - 2\rho^\theta \tau)\rho(-\frac{m}{\rho^2} + (1-2\tau)\theta\rho^{\theta-1})^2 d\tau$$

$$= I_1 + I_2 + I_3 + I_4,$$

$$(8.2.30)$$

where

$$I_1 = \frac{m^2}{\rho^3} \int_0^1 [\tau(1-\tau)]^\lambda g''(\frac{m}{\rho} + \rho^\theta - 2\rho^\theta \tau) d\tau, \qquad (8.2.31)$$

$$I_2 = \theta^2 \rho^{2\theta-1} \int_0^1 [\tau(1-\tau)]^\lambda g''(\frac{m}{\rho} + \rho^\theta - 2\rho^\theta \tau)(1-2\tau)^2 d\tau, \quad (8.2.32)$$

$$I_3 = (\theta + \theta^2)\rho^{\theta-1} \int_0^1 [\tau(1-\tau)]^\lambda (1-2\tau) g'(\frac{m}{\rho} + \rho^\theta - 2\rho^\theta \tau) d\tau, \qquad (8.2.33)$$

$$I_4 = 2\theta u \rho^{\theta-1} \int_0^1 [\tau(1-\tau)]^\lambda (1-2\tau) g''(\frac{m}{\rho} + \rho^\theta - 2\rho^\theta \tau) d\tau; \quad (8.2.34)$$

$$\eta_{\rho m}(\rho, m)$$

$$= \int_0^1 [\tau(1-\tau)]^\lambda g''(\frac{m}{\rho} + \rho^\theta - 2\rho^\theta \tau)(-\frac{m}{\rho^2} + (1-2\tau)\theta\rho^{\theta-1}) d\tau$$

$$= I_5 + I_6,$$

$$(8.2.35)$$

where

$$I_5 = -\frac{m}{\rho^2} \int_0^1 [\tau(1-\tau)]^\lambda g''(\frac{m}{\rho} + \rho^\theta - 2\rho^\theta \tau) d\tau, \qquad (8.2.36)$$

$$I_6 = \theta \rho^{\theta-1} \int_0^1 [\tau(1-\tau)]^\lambda (1-2\tau) g''(\frac{m}{\rho} + \rho^\theta - 2\rho^\theta \tau) d\tau; \quad (8.2.37)$$

and

$$\eta_{mm}(\rho, m) = \frac{1}{\rho} \int_0^1 [\tau(1-\tau)]^\lambda g''(\frac{m}{\rho} + \rho^\theta - 2\rho^\theta \tau) d\tau = I_7. \quad (8.2.38)$$

Since $2\theta - 1 = \gamma - 2$, then using (8.2.27), we have that $\varepsilon I_2(\rho_x)^2$ is bounded in $L^1_{loc}(R \times R^+)$.

Let $l(\tau) = \int_0^\tau [s(1-s)]^\lambda (1-2s) ds$. Then it is easy to see that $l(0) = l(1) = 0$ and hence

$$\rho^{\theta-1} \int_0^1 [\tau(1-\tau)]^\lambda (1-2\tau) g'(\frac{m}{\rho} + \rho^\theta - 2\rho^\theta \tau) d\tau$$
$$= 2\rho^{2\theta-1} \int_0^1 l(\tau) g''(\frac{m}{\rho} + \rho^\theta - 2\rho^\theta \tau) d\tau, \quad (8.2.39)$$

$$\rho^{\theta-1} \int_0^1 [\tau(1-\tau)]^\lambda (1-2\tau) g''(\frac{m}{\rho} + \rho^\theta - 2\rho^\theta \tau) d\tau$$
$$= 2\rho^{2\theta-1} \int_0^1 l(\tau) g'''(\frac{m}{\rho} + \rho^\theta - 2\rho^\theta \tau) d\tau. \quad (8.2.40)$$

Therefore, $\varepsilon I_3(\rho_x)^2$ and $\varepsilon I_4(\rho_x)^2$ are bounded in $L^1_{loc}(R \times R^+)$.

About I_6, we have

$$\varepsilon I_6 \rho_x m_x = \varepsilon u I_6 (\rho_x)^2 + \varepsilon I_6 \rho \rho_x u_x. \quad (8.2.41)$$

From (8.2.27), it is easy to see that the first term in the right-hand side of (8.2.41) is bounded in $L^1_{loc}(R \times R^+)$, and the second is also bounded in $L^1_{loc}(R \times R^+)$, since

$$|\varepsilon I_6 \rho \rho_x u_x|$$

$$= |\varepsilon \theta \rho^\theta \int_0^1 [\tau(1-\tau)]^\lambda (1-2\tau) g''(\frac{m}{\rho} + \rho^\theta - 2\rho^\theta \tau) d\tau \rho_x u_x|$$

$$\leq \varepsilon \theta (\rho^{2\theta-1}(\rho_x)^2 + \rho(u_x)^2)$$

$$\times |\int_0^1 [\tau(1-\tau)]^\lambda (1-2\tau) g''(\frac{m}{\rho} + \rho^\theta - 2\rho^\theta \tau) d\tau|.$$

Moreover

$$\varepsilon(I_1(\rho_x)^2 + 2I_5\rho_x m_x + I_7(m_x)^2)$$

is clearly bounded in $L^1_{loc}(R \times R^+)$ from the second estimate in (8.2.27). Thus we get the proof of Lemma 8.2.2. ∎

Lemma 8.2.3

$$\int\int_K (\varepsilon\rho_x^\varepsilon)^2 dxdt \to 0 \ \ as \ \varepsilon \to 0,$$

where K *is any bounded set in* $R \times R^+$.

Proof. We omit the superindex ε. Let $\Omega_1 = \{(x,t) \in R \times R^+ : \rho < \delta\}$ and $\Omega_2 = \{(x,t) \in R \times R^+ : \rho \geq \delta\}$. Then for a fixed constant $\delta \in (0,1)$, Ω_1 is an open set in $R \times R^+$. Let K_1 be any compact set in Ω_1 and choose $\phi \in C_0^2(\Omega_1)$ with a compact support set $S \subset \Omega_1$ and $\phi = 1$ on K_1, $0 \leq \phi \leq 1$.

Multiplying the first equation in (8.0.15) by 2ρ, we have

$$(\rho^2)_t + 2(\rho^2 u)_x - 2\rho u\rho_x = \varepsilon(\rho^2)_{xx} - 2\varepsilon(\rho_x)^2. \tag{8.2.42}$$

Multiplying (8.2.42) by the function ϕ and then integrating in $R \times R^+$, we get

$$\int_0^\infty \int_{-\infty}^\infty 2\varepsilon(\rho_x)^2 \phi dxdt$$

$$= \int_0^\infty \int_{-\infty}^\infty (\rho^2\phi_t + 2u\rho^2\phi_x + \varepsilon\rho^2\phi_{xx} + 2\rho u\rho_x\phi)dxdt \tag{8.2.43}$$

$$\leq C\delta^2 + C\delta\left(\int_0^\infty \int_{-\infty}^\infty (\rho_x)^2\phi dxdt\right)^{\frac{1}{2}},$$

where C is a suitable positive constant independent of ε, δ.

Thus from (8.2.43) we have

$$\int_0^\infty \int_{-\infty}^\infty \varepsilon^2(\rho_x)^2\phi dxdt = \int\int_S \varepsilon^2(\rho_x)^2\phi dxdt \leq C\delta^2, \tag{8.2.44}$$

or

$$\int\int_{K_1} \varepsilon^2 (\rho_x)^2 dx dt \leq C\delta^2. \tag{8.2.45}$$

Since $\varepsilon \rho^{\gamma-2}(\rho_x)^2$ s bounded in $L^1_{loc}(R \times R^+)$, then for fixed δ, we have

$$\int\int_{K_2} (\varepsilon \rho_x^\varepsilon)^2 dx dt \to 0 \text{ as } \varepsilon \to 0, \tag{8.2.46}$$

where K_2 is any bounded set in Ω_2. Combining (8.2.45) and (8.2.46) yields the proof of Lemma 8.2.3. ∎

Theorem 8.2.4 *For any weak entropy pair (η, q) of system (8.0.1), $\eta(\rho^\varepsilon, m^\varepsilon)_t + q(\rho^\varepsilon, m^\varepsilon)_x$ is compact in $H^{-1}_{loc}(R \times R^+)$, with respect to the viscosity solutions $(\rho^\varepsilon(x,t), m^\varepsilon(x,t))$.*

Proof. Multiplying system (8.0.15) by (η_ρ, η_m), we have

$$\eta(\rho^\varepsilon, m^\varepsilon)_t + q(\rho^\varepsilon, m^\varepsilon)_x$$
$$= \varepsilon \eta(\rho^\varepsilon, m^\varepsilon)_{xx} - \varepsilon(\rho_x^\varepsilon, m_x^\varepsilon) \cdot \nabla^2 \eta(\rho^\varepsilon, m^\varepsilon) \cdot (\rho_x^\varepsilon, m_x^\varepsilon)^T \tag{8.2.47}$$
$$= I_1 + I_2.$$

By the boundedness of the viscosity solutions $(\rho^\varepsilon, u^\varepsilon)$, the left-hand side of (8.2.47) is bounded in $W^{-1,\infty}$; by Lemma 8.2.2, I_2 is bounded in $L^1_{loc}(R \times R^+)$ and by Lemma 8.2.3, I_1 is compact in $H^{-1}_{loc}(R \times R^+)$ since

$$|\eta(\rho, m)_x| = |\eta_\rho \rho_x + \eta_m m_x| \leq C(|\rho_x| + (\rho)^{\frac{1}{2}} |u_x|).$$

Therefore, the proof of Theorem 8.2.4 is completed by Theorem 2.3.2. ∎

8.3   The Case of $\gamma > 3$

Theorem 8.3.1 *Let $\gamma > 3$ and $\rho^\varepsilon, u^\varepsilon$ be viscosity solutions given by (8.0.15)-(8.0.16). Then a subsequence of ρ^ε (still labelled ρ^ε) converges pointwisely to ρ, and (a subsequence of) u^ε converges pointwisely to u on the set $\{\rho(x,t) > 0\}$. In particular, $\rho^\varepsilon u^\varepsilon$ converges pointwisely to ρu.*

Proof. Taking two smooth functions $g(\xi_1), h(\xi_2)$ in (8.2.9)-(8.2.10) and using Theorem 2.1.4 and Theorem 2.3.2, we get

$$\int g(\xi_1)\overline{G(\xi_1)}d\xi_1 \int h(\xi_2)\overline{[\theta\xi_2 + (1-\theta)u]G(\xi_2)}d\xi_2$$

$$- \int h(\xi_2)\overline{G(\xi_2)}d\xi_2 \int g(\xi_1)\overline{[\theta\xi_1 + (1-\theta)u]G(\xi_1)}d\xi_1$$

$$= \int g(\xi_1)h(\xi_2)\overline{G(\xi_1)[\theta\xi_2 + (1-\theta)u]G(\xi_2)}d\xi_1 d\xi_2$$

$$- \int g(\xi_1)h(\xi_2)\overline{G(\xi_1)[\theta\xi_1 + (1-\theta)u]G(\xi_2)}d\xi_1 d\xi_2.$$

$$(8.3.1)$$

The last equality holds for arbitrary functions g, h, and this yields

$$\overline{G(\xi_1)}\,\overline{[\theta\xi_2 + (1-\theta)u]G(\xi_2)} - \overline{G(\xi_2)}\,\overline{[\theta\xi_1 + (1-\theta)u]G(\xi_1)}$$

$$= \overline{G(\xi_1)[\theta\xi_2 + (1-\theta)u]G(\xi_2)} - \overline{G(\xi_2)[\theta\xi_1 + (1-\theta)u]G(\xi_1)}$$

$$= \theta(\xi_2 - \xi_1)\overline{G(\xi_1)G(\xi_2)}.$$

$$(8.3.2)$$

Here and below we use the overbar to indicate the usual integration with respect to the Young measure; for instance

$$\overline{G(\xi)} = \int G(\rho, u - \xi)d\nu_{x,t}(\rho, u).$$

We may rewrite (8.3.2) as

$$\frac{\theta}{1-\theta}\left[\frac{\overline{G(\xi_1)G(\xi_2)}}{\overline{G(\xi_1)}\,\overline{G(\xi_2)}} - 1\right] = \frac{1}{\xi_2 - \xi_1}\left[\frac{\overline{uG(\xi_2)}}{\overline{G(\xi_2)}} - \frac{\overline{uG(\xi_1)}}{\overline{G(\xi_1)}}\right], \qquad (8.3.3)$$

for $\xi_1, \xi_2 \in \zeta$, where ζ is any open connected component in the union of the open intervals $(u - \rho^\theta, u + \rho^\theta)$, for which $(\rho, u) \in$ supp ν.

The first step of the proof is to show that for $\gamma > 3$,

$$\frac{\overline{uG(\xi)}}{\overline{G(\xi)}} \qquad \text{is a nonincreasing function of } \xi \in \zeta. \qquad (8.3.4)$$

We let $f_0(\xi)$ abbreviate $f_0(\xi) = \frac{G(\xi)-\overline{G(\xi)}}{\overline{G(\xi)}}$, so that (8.3.3) takes the equivalent form

$$\frac{\theta}{1-\theta} \overline{f_0(\xi_1)f_0(\xi_2)} = \frac{1}{\xi_2-\xi_1}\left[\frac{\overline{uG(\xi_2)}}{\overline{G(\xi_2)}} - \frac{\overline{uG(\xi_1)}}{\overline{G(\xi_1)}}\right]. \qquad (8.3.5)$$

Sending ξ_2 to $\xi_1 = \xi$ in (8.3.5), we should end up with

$$\frac{\theta}{1-\theta} \overline{f_0^2(\xi)} = \frac{\partial}{\partial\xi}\left(\frac{\overline{uG(\xi)}}{\overline{G(\xi)}}\right), \qquad (8.3.6)$$

and (8.3.4) follows, since

$$\frac{\theta}{1-\theta} = \frac{\gamma-1}{3-\gamma} < 0,$$

and hence the left-hand side of (8.3.6) is nonpositive for $\gamma > 3$.

Notice that there is no difficulty passing to the limit on the right-hand side of (8.3.5) in the sense of distributions since $\overline{G(\xi)}$ does not vanish on ζ. In order to pass to the limit on the left-hand side in $L_{loc}^2(\zeta)$, we require $f_0(\xi)$ and hence $G(\xi) \in L^2(R_\xi)$; but since $\|G(\xi)\|_{L^2(R_\xi)}^2 = \rho^{\frac{5-\gamma}{2}}\int_{-1}^{1}(1-\tau^2)^{2\lambda}d\tau$, this requirement of $L^2(R_\xi)$-integrability restricts the range of γ with $\gamma < 5$. To extend the statement of (8.3.4) for all $\gamma > 3$, we first mollify both sides of (8.3.5) with a unit mass mollifier, $\psi_\alpha(\xi) \geq 0$, and denote $f_\alpha = f_0 * \psi_\alpha$. Then we have

$$\frac{\theta}{1-\theta} \overline{f_\alpha(\xi_1)f_\alpha(\xi_2)} = \frac{1}{\xi_2-\xi_1}\left[\frac{\overline{uG(\xi_2)}}{\overline{G(\xi_2)}} - \frac{\overline{uG(\xi_1)}}{\overline{G(\xi_1)}}\right] * \psi_\alpha(\xi_1) * \psi_\alpha(\xi_2).$$
$$(8.3.7)$$

Thanks to the boundedness of the left-hand side and the smoothness of the right-hand side, we may now take $\xi_2 = \xi_1 = \xi$, to find out that

$$\frac{\theta}{1-\theta} \overline{f_\alpha^2(\xi)} = \frac{1}{\xi_2-\xi_1}\left[\frac{\overline{uG(\xi_2)}}{\overline{G(\xi_2)}} - \frac{\overline{uG(\xi_1)}}{\overline{G(\xi_1)}}\right] * \psi_\alpha(\xi_1) * \psi_\alpha(\xi_2)\,|_{\xi_2=\xi_1=\xi}.$$
$$(8.3.8)$$

If we now let α tend to zero, then the left-hand side of (8.3.8) yields a negative measure since $1 - \theta < 0$, whereas the right-hand side tends to

$$\frac{\partial}{\partial\xi}\left(\frac{\overline{uG(\xi)}}{\overline{G(\xi)}}\right),$$

which recovers the desired (8.3.4).

To complete the proof of Theorem 8.3.1, we need the following two necessary lemmas. The first one is about the construction of the set ζ:

Lemma 8.3.2 *Let the set ζ be the union of all open intervals $(u_i - \rho_i^\theta, u_i + \rho_i^\theta)$ for all points $(\rho_i, u_i) \in \text{supp}\,\nu$. Then ζ must be open and connected.*

The connection in Lemma 8.3.2 is not obvious. For instance, if the support set of ν contains only two points (ρ_1, u_1) and (ρ_2, u_2) satisfying

$$u_1 + \rho_1^\theta < u_2 - \rho_2^\theta,$$

then clearly ζ consists of two disjoint open intervals $(u_1 - \rho_1^\theta, u_1 + \rho_1^\theta)$ and $(u_2 - \rho_2^\theta, u_2 + \rho_2^\theta)$.

Proof of Lemma 8.3.2. To prove Lemma 8.3.2, let the support set S of the Young measure ν determined by the sequence of viscosity solutions to the Cauchy problem (8.0.15), (8.0.16) be contained in the smallest characteristic triangle:

$$\Sigma_1 = \{(\rho, u) : w \le w_0, z \ge z_0, \rho \ge 0\}.$$

Then clearly $z_0 \le z \le w \le w_0$ (see Figure 8.2).

If the support set S of ν is not wholly contained in the vacuum line $\rho = 0$, then we now show that the intersection point P_0 of the lines $w = w_0$ and $z = z_0$ must be in S, i.e.,

$$P_0 \in \text{supp}\,\nu. \tag{8.3.9}$$

If this is done, then ζ must be the open interval (z_0, w_0).

For the polytropic case $P(\rho) = c\rho^\gamma$ with the exponent $\gamma > 1$ and $c = \frac{1}{\gamma}$, the entropy equation (8.2.5) is reduced to

$$\eta_{\rho\rho} = \rho^{\gamma-3}\eta_{uu}. \tag{8.3.10}$$

If k denotes a positive constant, then the function $\eta = h(\rho)e^{ku}$ solves (8.3.10) provided that

$$h''(\rho) = k^2\rho^{\gamma-3}h. \tag{8.3.11}$$

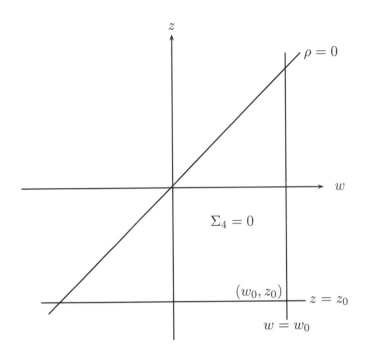

FIGURE 8.2

Let $a(\rho) = \rho^{\frac{1}{4}(3-\gamma)}, s = \frac{2k}{\gamma-1}\rho^{\frac{1}{2}(\gamma-1)}$. Then $h = a(\rho)\phi(s)$ solves (8.3.11) if and only if ϕ satisfies the standard Fuchsian equation

$$\phi'' - (1 + \mu s^2)\phi = 0, \qquad (8.3.12)$$

where $\mu = \frac{4-(\gamma-1)^2}{4(\gamma-1)^2} < 0$.

The second equation in (8.2.4) is

$$q_u = \rho\eta_\rho + u\eta_u. \qquad (8.3.13)$$

If

$$\eta_k = h(\rho)e^{ku}, \qquad (8.3.14)$$

then

$$(q_k)_u = \rho h'(\rho)e^{ku} + kuh(\rho)e^{ku}$$

and hence one entropy flux corresponding to η_k is

$$
\begin{aligned}
q_k &= uh(\rho)e^{ku} + (\rho h' - h)e^{ku}/k \\
&= \eta_k \left(u + \rho^{\frac{1}{2}(\gamma-1)} \frac{\phi'}{\phi} - \frac{\gamma+1}{4k} \right).
\end{aligned}
\tag{8.3.15}
$$

If

$$
\eta_{-k} = h(\rho)e^{-ku},
\tag{8.3.16}
$$

then

$$
(q_{-k})_u = \rho h'(\rho)e^{-ku} - kuh(\rho)e^{-ku}
$$

and hence one entropy flux corresponding to η_{-k} is

$$
\begin{aligned}
q_{-k} &= uh(\rho)e^{-ku} + (h - \rho h')e^{-ku}/k \\
&= \eta_{-k} \left(u - \rho^{\frac{1}{2}(\gamma-1)} \frac{\phi'}{\phi} + \frac{\gamma+1}{4k} \right).
\end{aligned}
\tag{8.3.17}
$$

Then two progressing waves of system (8.0.1) for the case of $P(\rho) = \frac{1}{\gamma}\rho^\gamma$ are provided by (8.3.14)-(8.3.15) and (8.3.16)-(8.3.17).

We may use the method of Frobenius again to obtain a series solution of Equation (8.3.12) as follows:

$$
\phi = s^j \sum_{n=0}^{\infty} e_n s^n,
\tag{8.3.18}
$$

where e_0 can be any positive constant, $j = \frac{\gamma+1}{2(\gamma-1)} > 0$ is the larger root of the equation $j(j-1) = \mu$ and e_n satisfy

$$
e_n = \frac{e_{n-1}}{(2n+j)(2n+j-1) - \mu}, \quad \text{for } n \geq 1.
\tag{8.3.19}
$$

It is easy to check that $\phi(s) > 0$ and $\phi'(s) > 0$.

Let $\eta_k = a(\rho)\phi(s)e^{ku}$. Then using Lemma 6.2.1, we have

$$
\eta_k = a(\rho)\phi(s)e^{-s}e^{kw} = e^{kw}(a(\rho) + O(\frac{1}{k}))
\tag{8.3.20}
$$

on $\rho > 0$ as k approaches infinity and the corresponding flux function is of the form

$$
\begin{aligned}
q_k &= \eta_k \left(u + \rho^{\frac{1}{2}(\gamma-1)} + \rho^{\frac{1}{2}(\gamma-1)} \left(\frac{\phi'(s)}{\phi(s)} - 1 \right) - \frac{\gamma+1}{4k} \right) \\
&= \eta_k (\lambda_2 - \frac{\gamma+1}{4k} + O(\frac{1}{k^2}))
\end{aligned}
\tag{8.3.21}
$$

on $\rho > 0$.

Similarly let $\eta_{-k} = a(\rho)\phi(s)e^{-ku}$. Then

$$\eta_{-k} = a(\rho)\phi(s)e^{-s}e^{-kz} = e^{-kz}(a(\rho) + O(\frac{1}{k})) \tag{8.3.22}$$

on $\rho > 0$ and the corresponding flux function

$$\begin{aligned}
q_{-k} &= \eta_{-k}\left(u - \rho^{\frac{1}{2}(\gamma-1)} - \rho^{\frac{1}{2}(\gamma-1)}(\frac{\phi'(s)}{\phi(s)} - 1) + \frac{\gamma+1}{4k}\right) \\
&= \eta_{-k}(\lambda_1 + \frac{\gamma+1}{4k} + O(\frac{1}{k^2}))
\end{aligned} \tag{8.3.23}$$

on $\rho > 0$. It is clear that the entropies $\eta_{\pm k}$ given in (8.3.20) and (8.3.22) are weak entropies, since they are zero at the vacuum line $\rho = 0$. Using these two positive progressing entropy pairs (8.3.20)-(8.3.21) and (8.3.22)-(8.3.23), we can complete the proof of (8.3.9). We shall argue by a contradiction. Suppose that the vertex P_0 does not lie in supp ν. Using $(\eta_{\pm k}, q_{\pm k})$ to the measure equation (6.4.2), we have

$$\frac{< \nu, q_{-k} >}{< \nu, \eta_{-k} >} - \frac{< \nu, q_k >}{< \nu, \eta_k >} = \frac{< \nu, \eta_k q_{-k} - \eta_{-k} q_k >}{< \nu, \eta_{-k} >< \nu, \eta_k >}. \tag{8.3.24}$$

Notice that

$$\eta_k q_{-k} - \eta_{-k} q_k = e^{k(w-z)}\{(\lambda_2 - \lambda_1)a(\rho) + O(\frac{1}{k})\} \tag{8.3.25}$$

and $(w_0, z_0) \notin S$. Then we have

$$| < \nu, \eta_k q_{-k} - \eta_{-k} q_k > | \le a_0 e^{k(w_0 - z_0 - \delta_0)} \tag{8.3.26}$$

for appropriate positive constants a_0, δ_0 and sufficiently large k. On the other hand,

$$| < \nu, \eta_k > | \ge a_1 e^{k(w_0 - \frac{\delta_0}{4})}, \quad | < \nu, \eta_{-k} > | \ge a_2 e^{-k(z_0 + \frac{\delta_0}{4})} \tag{8.3.27}$$

for appropriate positive constants a_1, a_2 since S is minimal. Therefore

$$\left| \frac{< \nu, \eta_k q_{-k} - \eta_{-k} q_k >}{< \nu, \eta_{-k} >< \nu, \eta_k >} \right| \le \frac{a_0}{a_1 a_2} e^{-\frac{k\delta_0}{2}} \to 0, \quad k \to \infty. \tag{8.3.28}$$

Similar to Section 6.4, we introduce two new probability measures $\mu_k^{\pm}$ on S defined by

$$< \mu_k^{\pm}, h >=< \nu, h\eta_{\pm k} > / < \nu, \eta_{\pm k} > \tag{8.3.29}$$

where $h = h(u, v)$ denotes an arbitrary continuous function. Clearly μ_k^+ and μ_k^- both are uniformly bounded with respect to k. Then as a consequence of weak-star compactness, there exist probability measures $\mu^\pm$ on S such that

$$< \mu^\pm, h > = \lim_{k \to \infty} < \mu_k^\pm, h >, \qquad (8.3.30)$$

after the selection of an appropriate subsequence.

Moreover, the measures μ^+, μ^- are respectively concentrated on the boundary sections of S associated with w and z, i.e.,

$$\text{supp } \mu^+ = S \bigcap \{(\rho, u) : w = w_0\} \qquad (8.3.31)$$

and

$$\text{supp } \mu^- = S \bigcap \{(\rho, u) : z = z_0\}. \qquad (8.3.32)$$

Then for the left-hand side of (8.3.24), there holds

$$\frac{< \nu, q_{-k} >}{< \nu, \eta_{-k} >} - \frac{< \nu, q_k >}{< \nu, \eta_k >} \to < \mu^-, \lambda_1 > - < \mu^+, \lambda_2 > \quad \text{as } k \to \infty. \qquad (8.3.33)$$

But

$$\lambda_{1w} = \lambda_{2z} = \frac{3 - \gamma}{4} > 0,$$

which implies that

$$\lambda_2(w_0, z) \geq \lambda_2(w_0, z_0) > \lambda_1(w_0, z_0) \geq \lambda_1(w, z_0)$$

and hence

$$< \mu^-, \lambda_1 > - < \mu^+, \lambda_2 > < 0,$$

which is in contradiction with (8.3.24) and (8.3.28). This completes the proof of (8.3.9) and hence that of Lemma 8.3.2. ∎

The second lemma is stated as:

Lemma 8.3.3 *Let $\zeta = (z_0, w_0)$ stand for the open connected component in Lemma 8.3.2, and let $u_0 = (z_0 + w_0)/2$. Then*

$$\lim_{\xi \to w_0} \overline{\frac{uG(\xi)}{G(\xi)}} \geq u_0, \quad \lim_{\xi \to z_0} \overline{\frac{uG(\xi)}{G(\xi)}} \leq u_0. \qquad (8.3.34)$$

Proof. According to (8.3.4), $\overline{uG}/G$ is a monotone function on ζ, and we turn to consider its one-side limits as $\xi \to w_0$ and $\xi \to z_0$. The values of (ρ, u) such that $G(\xi) > 0$ in an interval $(w_0 - \varepsilon, w_0)$ satisfy

$$u + \rho^\theta \geq w_0 - \varepsilon,$$

and therefore, since $w_0 \leq u - \rho^\theta$ for these (ρ, u) values, we have

$$\lim_{\xi \to w_0} \frac{\overline{uG(\xi)}}{G(\xi)} \geq \min\{u; (\rho, u) \in \operatorname{supp}\nu, u + \rho^\theta = w_0\}$$
$$\geq \frac{w_0 + z_0}{2}. \tag{8.3.35}$$

A similar argument holds for z_0, thus we finish the proof of Lemma 8.3.3. ∎

Now we are in the position to complete the proof of Theorem 8.3.1.

Combining (8.3.34) with (8.3.35) we obtain that $\overline{uG(\xi)}/G(\xi)$ is a constant, which in turn tells us, by (8.3.8), that $\overline{f_\alpha^2(\xi)} = 0$. Hence, $f_\alpha(\xi)$ vanishes on the support of ν and in particular, by letting $\alpha \to 0$, so does $f_0(\xi)$:

$$f_0(\xi) = \frac{G(\rho, u - \xi)}{G(\xi)} - 1 = 0, \quad (\rho, u) \in \operatorname{supp}\nu. \tag{8.3.36}$$

This shows that on the set $\{\rho > 0\}$, the Young measure ν is reduced to a Dirac mass and the conclusion holds as usual. This completes the proof of Theorem 8.3.1. ∎

8.4   The Case of $1 < \gamma \leq 3$

From the entropy-entropy flux equations (8.2.3) of system (8.0.1) and by simple calculations, we have the following four weak pairs of entropy-entropy flux to system (8.0.1):

$$(\eta_1, \ q_1) = (\rho, \ m), \tag{8.4.1}$$

$$(\eta_2, \ q_2) = (m, \ \frac{m^2}{\rho} + P(\rho)), \tag{8.4.2}$$

$$(\eta_3,\ q_3) = \Big(\frac{m^2}{2\rho} + \rho \int^\rho \frac{P(s)}{s^2} ds, \ \frac{m^3}{2\rho^2} + \big(\frac{P(\rho)}{\rho} + \int^\rho \frac{P(s)}{s^2} ds \big) m \Big),$$

$$(8.4.3)$$

$$(\eta_4, q_4) = \Big(\frac{m^3}{\rho^2} + 6m \int^\rho \frac{P(s)}{s^2} ds,$$

$$\frac{m^4}{\rho^3} + 3\big(\frac{P(\rho)}{\rho^2} + \frac{2}{\rho} \int^\rho \frac{P(s)}{s^2} ds \big) m^2$$

$$(8.4.4)$$

$$+ 6 \big(P(\rho) \int^\rho \frac{P(s)}{s^2} ds - \int^\rho \frac{P^2(s)}{s^2} ds \big) \Big).$$

Let

$$v = (\rho, m), \quad f(v) = \big(m, \frac{m^2}{\rho} + P(\rho) \big)^T,$$

$$\bar{v} = (\bar{\rho}, \bar{m}) = (< \nu, \rho >, < \nu, m >), \quad \bar{u} = \bar{m}/\bar{\rho}$$

and

$$\begin{cases} Q\eta = \eta(v) - \eta(\bar{v}) - \Delta\eta(\bar{v})(v - \bar{v}) \\ Q^\star q = q(v) - q(\bar{v}) - \Delta\eta(\bar{v})(f(v) - f(\bar{v})). \end{cases}$$

$$(8.4.5)$$

Then $(Q\eta_i, Q^\star q_i)(i = 1, 2, 3, 4)$ are also the entropy-entropy flux pairs of system (8.0.1) and satisfy the measure equation

$$\Big\langle \nu, \begin{vmatrix} Q\eta_i & Q^\star q_i \\ Q\eta_j & Q^\star q_j \end{vmatrix} \Big\rangle = \begin{vmatrix} < \nu, Q\eta_i > & < \nu, Q^\star q_i > \\ < \nu, Q\eta_j > & < \nu, Q^\star q_j > \end{vmatrix} \quad (i, j = 1, 2, 3, 4; i \ne j),$$

$$(8.4.6)$$

for the polytropic gas, i.e., if P takes the special form $P(\rho) = c\rho^\gamma$, where $\gamma > 1$.

For fixed (x, t), $\bar{v}$ is a scalar vector, thus it is clear that

$$Q\eta_1 = \rho - \bar{\rho}, \quad Q^\star q_1 = m - \bar{m}, \tag{8.4.7}$$

$$Q\eta_2 = m - \bar{m}, \quad Q^\star q_2 = \rho u^2 + P(\rho) - \bar{\rho}\bar{u}^2 - P(\bar{\rho}) \tag{8.4.8}$$

are also entropy-entropy flux pairs of system (8.0.1).

By simple calculations from (8.4.5), there hold:

$$Q\eta_3 = \frac{1}{2}\rho u^2 - \frac{1}{2}\bar{\rho}\bar{u}^2 + \frac{1}{2}\bar{u}^2(\rho - \bar{\rho}) - \bar{u}(m - \bar{m}) + \rho \int^{\rho} \frac{P(s)}{s^2} ds$$

$$-\bar{\rho}\int^{\bar{\rho}} \frac{P(s)}{s^2} ds - \left(\int^{\bar{\rho}} \frac{P(s)}{s^2} ds + \frac{P(\bar{\rho})}{\bar{\rho}}(\rho - \bar{\rho}) \right)$$

$$= \frac{1}{2}\rho(u - \bar{u})^2 + Q(\rho \int^{\rho} \frac{P(s)}{s^2} ds),$$

$$(8.4.9)$$

$$Q^\star q_3 = \frac{1}{2}m(u - \bar{u})^2 + (u - \bar{u})(P(\rho) - P(\bar{\rho}))$$

$$+u(\rho \int_{\bar{\rho}}^{\rho} \frac{P(s)}{s^2} ds - \frac{P(\bar{\rho})}{\bar{\rho}}(\rho - \bar{\rho})),$$

$$(8.4.10)$$

$$Q\eta_4 = 6m \int_{\bar{\rho}}^{\rho} \frac{P(s)}{s^2} ds + \rho(u - \bar{u})^2(u + 2\bar{u}) - \frac{6\bar{u}}{\bar{\rho}}P(\bar{\rho})(\rho - \bar{\rho}),$$

$$(8.4.11)$$

$$Q^\star q_4 = (6u^2\rho + 6P(\rho)) \int_{\bar{\rho}}^{\rho} \frac{P(s)}{s^2} ds - 6 \int_{\bar{\rho}}^{\rho} \frac{P^2(s)}{s^2} ds$$

$$+3P(\rho)(u^2 - \bar{u}^2) + u\rho(u - \bar{u})^2(u + 2\bar{u}) - \frac{6\bar{u}P(\bar{\rho})}{\bar{\rho}}(m - \bar{m}).$$

$$(8.4.12)$$

It follows from (8.4.6) that

$$< \nu, Q\eta_1 Q^\star q_2 - Q\eta_2 Q^\star q_1 >$$
$$=< \nu, (P(\rho) - P(\bar{\rho}))(\rho - \bar{\rho}) - (u - \bar{u})^2 \bar{\rho}\rho > \qquad (8.4.13)$$
$$= 0$$

and

$$< \nu, Q\eta_1 Q^\star q_3 - Q\eta_3 Q^\star q_1 >$$
$$=< \nu, (u - \bar{u})(\rho - \bar{\rho})P(\rho) - \bar{\rho}\rho(u - \bar{u}) \int_{\bar{\rho}}^{\rho} \frac{P(s)}{s^2} ds - \frac{1}{2}\bar{\rho}\rho(u - \bar{u})^3$$
$$= 0,$$

$$(8.4.14)$$

$$< \nu, Q\eta_1 Q^\star q_4 - Q\eta_4 Q^\star q_1 >$$

$$=< \nu, 3(\rho - \bar{\rho})(2Q_2 + (u^2 - \bar{u}^2)P(\rho)) - \bar{\rho}\rho(u - \bar{u})^3(u + 2\bar{u})$$

$$-6\bar{\rho}\rho u(u - \bar{u}) \int_{\bar{\rho}}^{\rho} \frac{P(s)}{s^2} ds >$$

$$= 0,$$

$$(8.4.15)$$

where

$$Q_2 = P(\rho) \int_{\bar{\rho}}^{\rho} \frac{P(s)}{s^2} ds - \int_{\bar{\rho}}^{\rho} \frac{P^2(s)}{s^2} ds. \qquad (8.4.16)$$

Calculate $(8.4.15) - 6\bar{u} \times (8.4.14)$. We have

$$< \nu, 3(\rho - \bar{\rho})(2Q_2 + (u - \bar{u})^2 P(\rho) - \bar{\rho}\rho(u - \bar{u})^4$$

$$-6\bar{\rho}\rho(u - \bar{u})^2 \int_{\bar{\rho}}^{\rho} \frac{P(s)}{s^2} ds > \qquad (8.4.17)$$

$$= 0.$$

It follows from (8.4.14) that

$$< \nu, \tfrac{1}{2}\bar{\rho}\rho(u - \bar{u})^3 >$$

$$< \nu, \ u - \bar{u})(\rho - \bar{\rho})P(\rho) - \bar{\rho}\rho(u - \bar{u}) \int_{\bar{\rho}}^{\rho} \frac{P(s)}{s^2} ds > . \qquad (8.4.18)$$

Using this and the measure equation

$$< \nu, Q\eta_2 Q^\star q_3 - Q\eta_3 Q^\star q_2 >$$

$$=< \nu, Q\eta_2 >< \nu, Q^\star q_3 > - < \nu, Q\eta_3 >< \nu, Q^\star q_2 >, \qquad (8.4.19)$$

we have

$$< \nu, \ \rho u^2 - \bar{\rho}\bar{u}^2 >< \nu, \ \tfrac{1}{2}\rho(u - \bar{u})^2 + \int_{\bar{\rho}}^{\rho} \frac{P(s)}{s^2} ds >$$

$$+ < \nu, P(\rho) >< \nu, Q_1 > - < \nu, P(\rho)Q_1 >$$

$$+ \frac{1}{2} < \nu, P(\rho) >< \nu, \rho(u - \bar{u})^2 > - \frac{1}{2} < \nu, \rho P(\rho)(u - \bar{u})^2 >$$

$$+ < \nu, \rho(u - \bar{u})^2(P(\rho) - P(\bar{\rho})) >$$

$$= 0,$$

$$(8.4.20)$$

where

$$Q_1 = \rho \int_{\bar\rho}^{\rho} \frac{P(s)}{s^2} ds - \frac{P(\bar\rho)}{\bar\rho}(\rho - \bar\rho).$$

By assumption $\rho \geq \rho_0 > 0$, then $\bar\rho \geq \rho_0 > 0$ also. Define:

$$O_{ij} = < \nu, O((u - \bar u)^i (\rho - \bar\rho)^j) >, \text{ where } O \text{ denotes an } L^\infty \text{ bound.}$$

By simple calculations, there hold

$$Q\eta_3 = \frac{1}{2}\bar\rho(u - \bar u)^2 + \frac{P'(\bar\rho)}{2\bar\rho}(\rho - \bar\rho)^2 + O_{21} + O_{03}, \qquad (8.4.21)$$

$$Q\eta_4 = \frac{6P(\bar\rho)}{\bar\rho}(u - \bar u)(\rho - \bar\rho) + \frac{3\bar u P'(\bar\rho)}{\bar\rho}(\rho - \bar\rho)^2$$
$$+ 3\bar u \bar\rho(u - \bar u)^2 + O_{21} + O_{30} + O_{12} + O_{03}. \qquad (8.4.22)$$

$$Q^\star q_3 = P'(\bar\rho)(u - \bar u)(\rho - \bar\rho) + \frac{\bar u P'(\bar\rho)}{2\bar\rho}(\rho - \bar\rho)^2$$
$$+ \frac{\bar u \bar\rho}{2}(u - \bar u)^2 + O_{21} + O_{12} + O_{30} + O_{03} \qquad (8.4.23)$$

and

$$Q^\star q_4 = \left(\frac{3P'(\bar\rho)P(\bar\rho)}{\bar\rho^2} + 3\bar u^2\frac{P'(\bar\rho)}{\bar\rho}\right)(\rho - \bar\rho)^2$$
$$+ \left(6\bar u P'(\bar\rho) + \frac{6\bar u P(\bar\rho)}{\bar\rho}\right)(u - \bar u)(\rho - \bar\rho) \qquad (8.4.24)$$
$$+ 3(\bar\rho\bar u^2 + P(\bar\rho))(u - \bar u)^2 + O_{30} + O_{21} + O_{12} + O_{03}.$$

It follows from (8.4.17) that

$$< \nu, \left(\frac{2P''(\bar\rho)P(\bar\rho) + (P'(\bar\rho))^2}{\bar\rho^2} - \frac{2P'(\bar\rho)P(\bar\rho)}{\bar\rho^3}\right)(\rho - \bar\rho)^4$$

$$+ \frac{3P'(\bar\rho)P(\bar\rho)}{\bar\rho^2}(\rho - \bar\rho)^3 - \bar\rho^2(u - \bar u)^4 - 3P(\bar\rho)(u - \bar u)^2(\rho - \bar\rho) >$$
$$+ O_{05} + O_{41} + O_{23} = 0, \qquad (8.4.25)$$

and from (8.4.20) that

$$(\frac{P'(\bar\rho)}{2} + \frac{\bar\rho}{4}P''(\bar\rho)) < \nu, (\rho - \bar\rho)^2 >< \nu, (u - \bar u)^2 >$$

$$+(\frac{P'(\bar\rho)}{2} + \frac{\bar\rho}{4}P''(\bar\rho)) < \nu, (\rho - \bar\rho)^2(u - \bar u)^2 >$$

$$+\frac{2(P'(\bar\rho))^2 - 5\bar\rho P'(\bar\rho)P''(\bar\rho)}{12\bar\rho^2} < \nu, (\rho - \bar\rho)^4 >$$

$$+\frac{\bar\rho^2}{2}(< \nu, (u - \bar u)^2 >)^2 + \frac{P''(\bar\rho)P'(\bar\rho)}{4\bar\rho}(< \nu, (\rho - \bar\rho)^2 >)^2$$

$$+\frac{1}{2}\bar\rho P'(\bar\rho) < \nu, (\rho - \bar\rho)(u - \bar u)^2 > -\frac{(P'(\bar\rho))^2}{2\bar\rho} < \nu, (\rho - \bar\rho)^3 >$$

$$+O_{21}O_{20} + O_{20}O_{03} + O_{02}O_{03} + O_{21}O_{02} + O_{05} + O_{23}$$

$$= 0.$$

$$(8.4.26)$$

Calculate (8.4.25) + (8.4.26) × $\frac{6P(\bar\rho)}{\bar\rho P'(\bar\rho)}$. Then

$$(\frac{2(P'(\bar\rho))^2 - P(\bar\rho)P''(\bar\rho)}{2\bar\rho^2} - \frac{P'(\bar\rho)P(\bar\rho)}{\bar\rho^3}) < \nu, (\rho - \bar\rho)^4 >$$

$$+\frac{3P(\bar\rho)P''(\bar\rho)}{2\bar\rho^2}(< \nu, (\rho - \bar\rho)^2 >)^2$$

$$+(\frac{3P(\bar\rho)}{\bar\rho} + \frac{3P(\bar\rho)P''(\bar\rho)}{2P'(\bar\rho)})(< \nu, (\rho - \bar\rho)^2 >< \nu, (u - \bar u)^2 >$$

$$+ < \nu, (\rho - \bar\rho)^2(u - \bar u)^2 >) + \frac{3\bar\rho P(\bar\rho)}{P'(\bar\rho)}(< \nu, (u - \bar u)^2 >)^2$$

$$= \bar\rho^2 < \nu, (u - \bar u)^4 > +O_{05} + O_{41} + O_{23}$$

$$+O_{21}O_{20} + O_{21}O_{02} + O_{20}O_{03} + O_{02}O_{03}.$$

$$(8.4.27)$$

By simple calculations,

$$< \nu, Q\eta_3 Q^\star q_4 - Q\eta_4 Q^\star q_3 >=< \nu, \frac{1}{2}\rho P(\rho)(u - \bar{u})^4$$

$$+\rho P(\bar{\rho})(u - \bar{u})^4 + 6Q_1 Q_2 - 3\rho(u - \bar{u})^2 \int_{\bar{\rho}}^{\rho} \frac{P^2(s)}{s^2}ds$$

$$-6\rho(u - \bar{u})^2 P(\bar{\rho}) \int_{\bar{\rho}}^{\rho} \frac{P(s)}{s^2}ds - 3(\rho - \bar{\rho})(u - \bar{u})^2 \frac{P(\bar{\rho})}{\bar{\rho}} P(\rho) >$$

$$=< \nu, \frac{3}{2}\bar{\rho} P(\bar{\rho})(u - \bar{u})^4 - \frac{3P(\bar{\rho})P'(\bar{\rho})}{\bar{\rho}}(u - \bar{u})^2(\rho - \bar{\rho})^2$$

$$+\frac{3P(\bar{\rho})(P'(\bar{\rho}))^3}{2\bar{\rho}^3}(\rho - \bar{\rho})^4 + O_{41} + O_{23} + O_{05}.$$

$$(8.4.28)$$

Then from (8.4.21)-(8.4.24) and the following measure equation:

$$< \nu, Q\eta_3 Q^\star q_4 - Q\eta_4 Q^\star q_3 >$$

$$=< \nu, Q\eta_3 >< \nu, Q^\star q_4 > - < \nu, Q\eta_4 >< \nu, Q^\star q_3 >,$$

we have

$$< \nu, \frac{3}{2}\bar{\rho} P(\bar{\rho})(u - \bar{u})^4 - \frac{3P(\bar{\rho})P'(\bar{\rho})}{\bar{\rho}}(u - \bar{u})^2(\rho - \bar{\rho})^2$$

$$+\frac{3P(\bar{\rho})(P'(\bar{\rho}))^2}{2\bar{\rho}^3}(\rho - \bar{\rho})^4 >$$

$$= \frac{3}{2}\bar{\rho} P(\bar{\rho})(< \nu, (u - \bar{u})^2 >)^2 + \frac{3(P'(\bar{\rho}))^2 P(\bar{\rho})}{2\bar{\rho}^3}(< \nu, (\rho - \bar{\rho})^2 >)^2$$

$$+\frac{3P'(\bar{\rho})P(\rho)}{\bar{\rho}} < \nu, (\rho - \bar{\rho})^2 >< \nu, (u - \bar{u})^2 >$$

$$-\frac{6P'(\bar{\rho})P(\bar{\rho})}{\bar{\rho}} < \nu, (\rho - \bar{\rho})(u - \bar{u}) > +Error,$$

$$(8.4.29)$$

where *Error* is the higher order error given by

$$\begin{aligned}
Error \ &= O_{41} + O_{23} + O_{05} + O_{20}O_{30} + O_{20}O_{21} + O_{20}O_{12} \\
&\quad + O_{20}O_{03} + O_{02}O_{30} + O_{02}O_{21} + O_{02}O_{12} \\
&\quad + O_{02}O_{03} + O_{11}O_{21} + O_{11}O_{12} + O_{11}O_{30} + O_{11}O_{03}.
\end{aligned}$$

At this moment, let $P(\rho) = k^2 \rho^\gamma$ for $1 < \gamma \le 3$. Then we have from
(8.4.27) and (8.4.29) that

$$\begin{aligned}
\frac{\gamma^2 - \gamma}{2} &k^4 \bar{\rho}^{2\gamma-4} < \nu, (\rho - \bar{\rho})^4 > + \frac{3}{2}(\gamma^2 - \gamma)k^4\bar{\rho}^{2\gamma-4}(< \nu, (\rho - \bar{\rho})^2 >)^2 \\
&+ \frac{3}{2}(\gamma + 1)k^2\bar{\rho}^{\gamma-1}(< \nu, (\rho - \bar{\rho})^2 >< \nu, (u - \bar{u})^2 > \\
&+ < \nu, (\rho - \bar{\rho})^2(u - \bar{u})^2 >) + \frac{3}{\gamma}\bar{\rho}^2(< \nu, (u - \bar{u})^2 >)^2 \\
&= \bar{\rho}^2 < \nu, (u - \bar{u})^4 > + O_{05} + O_{41} + O_{23} \\
&\quad + O_{21}O_{20} + O_{21}O_{02} + O_{20}O_{03} + O_{02}O_{03}
\end{aligned}$$

$$(8.4.30)$$

and

$$\begin{aligned}
< \nu, &\frac{3}{2}k^2\bar{\rho}^{\gamma+1}(u - \bar{u})^4 - 3\gamma k^4\bar{\rho}^{2\gamma-2}(u - \bar{u})^2(\rho - \bar{\rho})^2 \\
&= \frac{3}{2}k^2\rho^{\gamma+1}(< \nu, (u - \bar{u})^2 >)^2 + \frac{3}{2}k^6\gamma^2\bar{\rho}^{3\gamma-5}(< \nu, (\rho - \bar{\rho})^2 >)^2 \\
&+ 3\gamma k^4\bar{\rho}^{2\gamma-2} < \nu, (\rho - \bar{\rho})^2 >< \nu, (u - \bar{u})^2 > \\
&- 6\gamma k^4\bar{\rho}^{2\gamma-2} < \nu, (\rho - \bar{\rho})(u - \bar{u}) > + \frac{3}{2}k^6\gamma^2\bar{\rho}^{3\gamma-5}(\rho - \bar{\rho})^4 > \\
&+ Error.
\end{aligned}$$

$$(8.4.31)$$

Calculate (8.4.30) $\times \frac{2\gamma k^2\bar{\rho}^{\gamma-1}}{\gamma+1}$ + (8.4.31). We have

$$\begin{aligned}
&\frac{3 - \gamma}{2(\gamma + 1)}k^2\bar{\rho}^{\gamma+1} < \nu, (u - \bar{u})^4 > \\
&+ \frac{5\gamma + 1}{2(\gamma + 1)}\gamma^2 k^6\bar{\rho}^{3\gamma-5} < \nu, (\rho - \bar{\rho})^4 > \\
&+ \frac{3(3 - \gamma)}{2(\gamma + 1)}k^2\bar{\rho}^{\gamma+1}(< \nu, (u - \bar{u})^2 >)^2 \\
&+ \frac{3(\gamma - 3)}{2(\gamma + 1)}\gamma^2 k^6\bar{\rho}^{3\gamma-5}(< \nu, (\rho - \bar{\rho})^2 >)^2 \\
&+ 6\gamma k^4\bar{\rho}^{2\gamma-2}(< \nu, (u - \bar{u})(\rho - \bar{\rho}) >)^2 + Error \\
&= 0.
\end{aligned}$$

$$(8.4.32)$$

If $1 < \gamma < 3$, and noticing that

$$(< \nu, (\rho - \bar{\rho})^2 >)^2 \leq < \nu, (\rho - \bar{\rho})^4 >,$$

we have from (8.4.32) that

$$\frac{3 - \gamma}{2(\gamma + 1)} k^2 \bar{\rho}^{\gamma+1} < \nu, (u - \bar{u})^4 >$$

$$+ \frac{4(\gamma - 1)}{\gamma + 1} \gamma^2 k^6 \bar{\rho}^{3\gamma-5} < \nu, (\rho - \bar{\rho})^4 >$$

$$+ \frac{3(3 - \gamma)}{2(\gamma + 1)} k^2 \bar{\rho}^{\gamma+1} (< \nu, (u - \bar{u})^2 >)^2 \tag{8.4.33}$$

$$+ 6\gamma k^4 \bar{\rho}^{2\gamma-2} (< \nu, (u - \bar{u})(\rho - \bar{\rho}) >)^2 + Error$$

$$\leq 0,$$

which implies that

$$C_1 < \nu, (u - \bar{u})^4 > + C_2 < \nu, (\rho - \bar{\rho})^4 > \leq 0, \tag{8.4.34}$$

for two suitable positive constants C_1 and C_2, which depend only on $k, \bar{\rho}, \gamma$ when the support of the ν is small. Thus ν is a Dirac measure and the support point is $(\bar{u}, \bar{\rho})$.

If $\gamma = 3$, first it follows from (8.4.33) that ν is concentrated on the line $\rho = \bar{\rho}$ and then it is reduced to the point $(\bar{u}, \bar{\rho})$ since we have from (8.4.13) that

$$< \nu, (u - \bar{u})^2 > = \gamma k^2 \bar{\rho}^{\gamma-3} < \nu, (\rho - \bar{\rho})^2 > + O_{21} + O_{03}. \tag{8.4.35}$$

This completes the reduction of the Young measure to be a Dirac measure in the case of $1 < \gamma \leq 3$.

8.5   Application on Extended River Flow System

In this section, we shall use the compactness framework introduced in the previous sections to study the river flow equations, a shallow-water model describing the vertical depth ρ and mean velocity u by

$$\begin{cases} \rho_t + (\rho u)_x = 0, \\ (\rho u)_t + (\rho u^2 + P(\rho))_x + a(x)\rho + c\rho u|u| = 0, \end{cases} \tag{8.5.1}$$

with bounded measurable initial data

$$(\rho(x,0), u(x,0)) = (\rho_0(x), u_0(x)), \quad \rho_0(x) \geq 0, \tag{8.5.2}$$

where the function $a(x)$ corresponds physically to the slope of the topography, $c\rho|u|$ to a friction term and c is a nonnegative constant.

Consider the Cauchy problem for the related parabolic system

$$\begin{cases} \rho_t + (\rho u)_x = \varepsilon \rho_{xx}, \\ (\rho u)_t + (\rho u^2 + P(\rho))_x + a(x)\rho + c\rho u|u| = \varepsilon(\rho u)_{xx}, \end{cases} \tag{8.5.3}$$

with initial data

$$(\rho^\varepsilon(x,0), (\rho^\varepsilon u^\varepsilon)(x,0)) = (\rho_0^\varepsilon(x), \rho_0^\varepsilon(x)u_0^\varepsilon(x)), \tag{8.5.4}$$

where $(\rho_0^\varepsilon(x), \rho_0^\varepsilon(x)u_0^\varepsilon(x))$ are given by (8.0.17), hence satisfy (8.0.18), (8.0.20).

Similar to Theorem 8.0.1, we have the main result in this section:

Theorem 8.5.1 *Let*

(1) $|a(x)| \leq M$ and M be a nonnegative constant;

(2) the initial data $(\rho_0(x), u_0(x))$ be bounded measurable and $\rho_0(x) \geq 0$;

(3) $P(\rho) \in C^2(0, \infty), P'(\rho) > 0, 2P'(\rho) + \rho P''(\rho) \geq 0$ for $\rho > 0$ and

$$\int_d^\infty \frac{\sqrt{P'(\rho)}}{\rho} d\rho = \infty, \quad \int_0^d \frac{\sqrt{P'(\rho)}}{\rho} d\rho < \infty, \quad \forall d > 0.$$

Then for fixed $\varepsilon > 0$, the smooth viscosity solution $(\rho^\varepsilon, \rho^\varepsilon u^\varepsilon)$ of the Cauchy problem (8.5.3), (8.5.4) exists and satisfies

$$0 < c(\varepsilon, t) \leq \rho^\varepsilon(x, t) \leq M_2(t), \quad \|u^\varepsilon(x, t)\| \leq M_2(t), \tag{8.5.5}$$

where $M_2(t)$ is a positive finite function of $t \in (0, \infty)$, being independent of ε, and $c(\varepsilon, t)$ is a positive function, which could tend to zero as ε tends to zero or t tends to infinity.

Moreover, for the case of $P(\rho) = k^2 \rho^\gamma, \gamma > 1$, there exists a subsequence (still labelled) $(\rho^\varepsilon(x,t), \rho^\varepsilon(x,t)u^\varepsilon(x,t))$ which converges almost everywhere on any bounded and open set $\Omega \subset R \times R^+$:

$$(\rho^\varepsilon(x,t), \rho^\varepsilon(x,t)u^\varepsilon(x,t)) \to (\rho(x,t), \rho(x,t)u(x,t)), \quad as\ \varepsilon \downarrow 0^+, \quad (8.5.6)$$

where the limit pair of functions $(\rho(x,t), \rho(x,t)u(x,t))$ is a weak solution of the Cauchy problem (8.5.1), (8.5.2).

Proof. Exactly analogous to the proof of Theorem 8.0.1 given in Sections 8.1-8.4, we can prove Theorem 8.5.1 if we have the L^∞ estimates (8.5.5).

Multiplying system (8.5.3) by (w_ρ, w_m) and $(\bar{z}_\rho, \bar{z}_m)$, respectively, where w, z $(\bar{z} = -z)$ are two Riemann invariants given in (8.0.9), we have

$$w_t + \lambda_2 w_x + a(x) + \frac{c|u|}{2}(w - \bar{z})$$

$$= \varepsilon w_{xx} + \frac{2\varepsilon}{\rho}\rho_x w_x - \frac{\varepsilon}{2\rho^2\sqrt{P'(\rho)}}(2P' + \rho P'')\rho_x^2 \quad (8.5.7)$$

$$\leq \varepsilon w_{xx} + \frac{2\varepsilon}{\rho}\rho_x w_x$$

and

$$\bar{z}_t + \lambda_1 \bar{z}_x - a(x) + \frac{c|u|}{2}(\bar{z} - w)$$

$$= \varepsilon \bar{z}_{xx} - \frac{2\varepsilon}{\rho}\rho_x \bar{z}_x - \frac{\varepsilon}{2\rho^2\sqrt{P'(\rho)}}(2P' + \rho P'')\rho_x^2 \quad (8.5.8)$$

$$\leq \varepsilon \bar{z}_{xx} - \frac{2\varepsilon}{\rho}\rho_x \bar{z}_x.$$

Making a transformation

$$w = X + Mt, \qquad \bar{z} = Y + Mt,$$

where M is the bound of $a(x)$, we have from (8.5.7)-(8.5.8) that

$$\begin{cases} X_t + \lambda_2 X_x + \dfrac{c|u|}{2}(X - Y) \leq \varepsilon X_{xx} + \dfrac{2\varepsilon}{\rho}\rho_x X_x, \\[3mm] Y_t + \lambda_1 Y_x + \dfrac{c|u|}{2}(Y - X) \leq \varepsilon Y_{xx} - \dfrac{2\varepsilon}{\rho}\rho_x Y_x, \end{cases} \quad (8.5.9)$$

with

$$X|_{t=0} = w|_{t=0} \le M_1, \ Y|_{t=0} = \bar{z}|_{t=0} \le M_1, \tag{8.5.10}$$

where M_1 is a positive constant depending only on the bound of the initial data of $\rho_0(x), u_0(x)$.

In the following we use the maximum principle to (8.5.9)-(8.5.10) to get the estimates

$$X(x,t) \le M_1, \qquad Y(x,t) \le M_1, \text{ for } (x,t) \in R \times [0,T], \tag{8.5.11}$$

which implies

$$w \le M_1 + Mt, \qquad z \ge -M_1 - Mt, \text{ for } (x,t) \in R \times [0,T] \tag{8.5.12}$$

and hence, the upper bound estimates of ρ^ε and u^ε.

To prove (8.5.11), we make the following transformation:

$$\bar{X} = X - M_1 - \frac{N(x^2 + CLe^t)}{L^2},$$
$$\bar{Y} = Y - M_1 - \frac{N(x^2 + CLe^t)}{L^2}, \tag{8.5.13}$$

where C, L are positive constants and N is the upper bound of X, Y on $R \times [0,T]$ (by local solution, N exists). From (8.5.9), it is easy to see that functions $\bar{X}$ and $\bar{Y}$ satisfy the inequalities

$$\bar{X}_t + \lambda_2 \bar{X}_x + \frac{c|u|}{2}(\bar{X} - \bar{Y}) + (CLe^t + 2\lambda_2 x - 2\varepsilon - \frac{4\varepsilon}{\rho}\rho_x x)\frac{N}{L^2}$$
$$\le \varepsilon \bar{X}_{xx} + \frac{2\varepsilon}{\rho}\rho_x \bar{X}_x \tag{8.5.14}$$

and

$$\bar{Y}_t + \lambda_1 \bar{Y}_x + \frac{c|u|}{2}(\bar{Y} - \bar{X}) + (CLe^t + 2\lambda_1 x - 2\varepsilon + \frac{4\varepsilon}{\rho}\rho_x x)\frac{N}{L^2}$$
$$\le \varepsilon \bar{Y}_{xx} - \frac{2\varepsilon}{\rho}\rho_x \bar{Y}_x. \tag{8.5.15}$$

Moreover,

$$\bar{X}_0(x) = X_0(x) - M_1 - \frac{CLN}{L^2} < 0,$$
$$\bar{Y}_0(x) = Y_0(x) - M_1 - \frac{CLN}{L^2} < 0 \tag{8.5.16}$$

and

$$\bar{X}(\pm L, t) < 0, \quad \bar{Y}(\pm L, t) < 0 \qquad (8.5.17)$$

since N is the upper bound of X, Y.

We have from (8.5.14)-(8.5.17) that

$$\bar{X}(x, t) < 0, \quad \bar{Y}(x, t) < 0, \text{ on } (-L, L) \times (0, T). \qquad (8.5.18)$$

We argue by assuming that (8.5.18) is violated for $\bar{X}$ (or $\bar{Y}$) at a point (x, t) in $(-L, L) \times (0, T)$. Let $\bar{t}$ be the least upper bound of values of t at which $\bar{X} < 0$. Then by the continuity we see that $\bar{X} = 0$, $\bar{Y} \leq 0$ at some points $(\bar{x}, \bar{t}) \in (-L, L) \times (0, T)$. So $\bar{X}_t \geq 0$, $\bar{X}_x = 0$ and $-\bar{X}_{xx} \geq 0$ at $(\bar{x}, \bar{t})$, i.e.,

$$\bar{X}_t + \lambda_2 \bar{X}_x - \varepsilon \bar{X}_{xx} - \frac{2\varepsilon}{\rho} \rho_x \bar{X}_x \geq 0 \text{ at } (\bar{x}, \bar{t}). \qquad (8.5.19)$$

But from the behaviors of local solution ρ^ε, we can choose C large enough so that

$$CLe^t + 2\lambda_2 x - 2\varepsilon - \frac{4\varepsilon}{\rho} \rho_x x > 0, \text{ on } (-L, L) \times (0, T), \qquad (8.5.20)$$

or

$$CLe^t + 2\lambda_1 x - 2\varepsilon + \frac{4\varepsilon}{\rho} \rho_x x > 0, \text{ on } (-L, L) \times (0, T). \qquad (8.5.21)$$

(8.5.19) and (8.5.20) yield a conclusion contradicting (8.5.14). So (8.5.18) is proved. Therefore for any point (x_0, t_0) in $(-L, L) \times (0, T)$,

$$\begin{aligned} X(x_0, t_0) &< M_1 + \frac{N(x_0^2 + CLe^{t_0})}{L^2}, \\ Y(x_0, t_0) &< M_1 + \frac{N(x_0^2 + CLe^{t_0})}{L^2}, \end{aligned} \qquad (8.5.22)$$

which yields the desired estimates in (8.5.11) if we let $L \uparrow \infty$ in (8.5.22), and hence completes the proof of Theorem 8.5.1. ∎

8.6 Related Results

The study of the existence of global weak solutions for the Cauchy problem (8.0.1), (8.0.2) has a long history. For the polytropic gas, the first large data existence theorem with locally finite total variation, for $\gamma = 1$ was obtained by Nishida [Ni], and for $\gamma \in (1, 1 + \delta)$, δ small, obtained by Smoller and Nishida [NS]. The method used in [Ni, NS] is called the Glimm's scheme method (cf. [Gl]).

Using the compensated compactness ideas developed by Tartar and Murat [Ta, Mu], DiPerna [Di2] established a global existence theorem for $\gamma = 1 + \frac{2}{N}, N \geq 5$ odd, with the aid of the viscosity method. Ding, Chen, and Luo [DCL1] and Chen [Ch1] proved the convergence of the Lax-Friedrichs scheme and the existence of global solutions with L^∞ large initial data with adiabatic exponent $\gamma \in (1, \frac{5}{3}]$. Lions, Perthame and Tadmor [LPT] proved the global existence of a weak solution for $\gamma \geq 3$ by applying the kinetic formulation coupled with the compensated compactness method. Finally, the method in [LPT] was successfully extended by Lions, Perthame and Souganidis [LPS] to fill up the gap $\gamma \in (\frac{5}{3}, 3)$ as well as a new proof for whole $\gamma > 1$. So the existence of a generalized solution for the Cauchy problem (8.0.1), (8.0.2) was completely resolved for the case of a polytropic gas.

The global smooth solution of the Cauchy problem (8.0.1), (8.0.2) for a class of smooth initial data with the vacuum for a general pressure $P(\rho)$ was obtained in [Lu5].

The global existence of L^∞ entropy solutions for system (8.0.1) with a special pressure $P(\rho)$ and arbitrarily large L^∞ initial data was established in [CL].

In this chapter, the proof for the case of $\gamma > 3$ comes from [LPT]. However, to avoid the use of many knotty mathematical formulas, we have not introduced the proofs given in [Di2, DCL1, Ch1, LPS] for the case of $\gamma \in (1, 3]$. Instead, in Section 8.4 we adopted another proof given by [CL2], although it needs some extra assumptions on the viscosity solutions.

The proof of Theorem 8.5.1 is from [KL2]. The related results about system (8.0.1) with inhomogeneous terms can be found in [DCL2, CG].

Chapter 9

Two Special Systems of Euler Equations

In this chapter, we consider the existence of global weak solutions for the following nonlinear hyperbolic systems:

$$\begin{cases} \rho_t + (\rho u)_x = 0 \\ u_t + (\frac{1}{2}u^2 + \int_0^\rho \frac{P'(s)}{s}ds)_x = 0, \end{cases} \qquad (9.0.1)$$

with bounded measurable initial data

$$(\rho(x,0), u(x,0)) = (\rho_0(x), u_0(x)), \quad \rho_0(x) \geq 0. \qquad (9.0.2)$$

For smooth solutions, system (9.0.1) is equivalent to the isentropic equations of gas dynamics (8.0.1), but these two systems are different for solutions with shock waves.

System (9.0.1) was first derived by S. Earnshaw for isentropic flow (cf. [Ea], [Wh]) and is also called the Euler equations of one-dimensional, compressible fluid flow (cf. [KM]). It is a scaling limit system of a Newtonian dynamics with long-range interaction for a continuous distribution of mass in R (cf. [Oe1, Oe2]) and also a hydrodynamic limit for the Vlasov equation (cf. [CEMP]).

Using the method introduced in Chapters 6 and 7, in this chapter, we shall study two special cases for $P(\rho)$:

$$P(\rho) = \int_0^\rho s^2 e^s ds, \text{ and } P(\rho) = \int_0^\rho s^2(s+d)^{\gamma-3}ds, \qquad (9.0.3)$$

121

where $\gamma > 3$.

When $P(\rho) = \int_0^\rho s^2 e^s ds$, let F be the mapping from R^2 into R^2 defined by

$$F : (\rho, u) \rightarrow \left(\rho u, \frac{1}{2}u^2 + \int_0^\rho s e^s ds\right).$$

Then two eigenvalues of dF are

$$\lambda_1 = u - \rho e^{\frac{\rho}{2}}, \quad \lambda_2 = u + \rho e^{\frac{\rho}{2}}, \tag{9.0.4}$$

and the corresponding right eigenvectors are

$$r_1 = (1, -e^{\frac{\rho}{2}})^T, \quad r_2 = (1, e^{\frac{\rho}{2}})^T. \tag{9.0.5}$$

By simple calculations,

$$\nabla \lambda_1 \cdot r_1 = -2e^{\frac{\rho}{2}} - \frac{\rho}{2}e^{\frac{\rho}{2}} < 0, \text{ for } \rho \geq 0, \tag{9.0.6}$$

and

$$\nabla \lambda_2 \cdot r_2 = 2e^{\frac{\rho}{2}} + \frac{\rho}{2}e^{\frac{\rho}{2}} > 0, \text{ for } \rho \geq 0. \tag{9.0.7}$$

Therefore, it follows from (9.0.4) that $\lambda_1 = \lambda_2$ at the line $\rho = 0$ at which the strict hyperbolicity fails to hold, and from (9.0.6)-(9.0.7) that both characteristic fields are genuinely nonlinear in the range $\rho \geq 0$.

Two Riemann invariants of system (9.0.1) for $P(\rho) = \int_0^\rho s^2 e^s ds$ are

$$z = u - 2e^{\frac{\rho}{2}}, \quad w = u + 2e^{\frac{\rho}{2}}. \tag{9.0.8}$$

Then

$$\lambda_1 = \frac{w+z}{2} - \frac{w-z}{2}\ln(\frac{w-z}{4}), \quad \lambda_2 = \frac{w+z}{2} + \frac{w-z}{2}\ln(\frac{w-z}{4}), \tag{9.0.9}$$

and

$$\lambda_{1w} = -\frac{1}{2}\ln(\frac{w-z}{4}), \quad \lambda_{2z} = -\frac{1}{2}\ln(\frac{w-z}{4}). \tag{9.0.10}$$

Just as given in (6.2.6) or (8.2.16), the entropies for any 2×2 hyperbolic conservation laws satisfy the equation

$$\eta_{wz} + \frac{\lambda_{2z}}{\lambda_2 - \lambda_1}\eta_w - \frac{\lambda_{1w}}{\lambda_2 - \lambda_1}\eta_z = 0. \tag{9.0.11}$$

Then the entropy equation of system (9.0.1) for the case of $P(\rho) = \int_0^\rho s^2 e^s ds$ is reduced to

$$\eta_{wz} - \frac{1}{2(w-z)}\eta_w + \frac{1}{2(w-z)}\eta_z = 0. \tag{9.0.12}$$

We recall that the entropies for the system of isentropic gas dynamics (8.0.1) satisfy

$$\eta_{wz} + \frac{c}{w-z}\eta_w - \frac{c}{w-z}\eta_z = 0, \tag{9.0.13}$$

where $c = \frac{3-\gamma}{2(\gamma-1)}$. So in a sense, the case of $P(\rho) = \int_0^\rho s^2 e^s ds$ is corresponding to the case of $\gamma = \infty$.

When $P(\rho) = \int_0^\rho s^2(s+d)^{\gamma-3}ds$, $\gamma > 3$, let F be the mapping from R^2 into R^2 defined by

$$F : (\rho, u) \to (\rho u, \frac{1}{2}u^2 + \int_0^\rho s(s+d)^{\gamma-3}ds).$$

Then two eigenvalues of dF are

$$\lambda_1 = u - \rho(\rho+d)^{\frac{1}{2}(\gamma-3)}, \quad \lambda_2 = u + \rho(\rho+d)^{\frac{1}{2}(\gamma-3)} \tag{9.0.14}$$

with corresponding right eigenvectors

$$r_1 = (1, -(\rho+d)^{\frac{1}{2}(\gamma-3)})^T, \quad r_2 = (1, (\rho+d)^{\frac{1}{2}(\gamma-3)})^T. \tag{9.0.15}$$

The two corresponding Riemann invariants for this case are

$$w = u + \frac{1}{2}(\gamma-1)(\rho+d)^{\frac{1}{2}(\gamma-1)}, \quad z = u - \frac{1}{2}(\gamma-1)(\rho+d)^{\frac{1}{2}(\gamma-1)}. \tag{9.0.16}$$

By simple calculations,

$$\nabla\lambda_1 \cdot r_1 = -2(\rho+d)^{\frac{1}{2}(\gamma-3)} - \frac{1}{2}(\gamma-3)\rho(\rho+d)^{\frac{1}{2}(\gamma-5)} < 0, \text{ for } \rho \geq 0, \tag{9.0.17}$$

and

$$\nabla\lambda_2 \cdot r_2 = 2(\rho+d)^{\frac{1}{2}(\gamma-3)} + \frac{1}{2}(\gamma-3)\rho(\rho+d)^{\frac{1}{2}(\gamma-5)} > 0, \text{ for } \rho \geq 0. \tag{9.0.18}$$

Therefore, it follows from (9.0.14) that $\lambda_1 = \lambda_2$ at the line $\rho = 0$ at which the strict hyperbolicity fails to hold, and from (9.0.17)-(9.0.18) that both characteristic fields are genuinely nonlinear in the range $\rho \geq 0$.

Consider the Cauchy problem for the related parabolic system

$$\begin{cases} \rho_t + (\rho u)_x = \varepsilon \rho_{xx} \\ u_t + (\frac{1}{2}u^2 + \int_0^\rho \frac{P'(s)}{s} ds)_x = \varepsilon u_{xx}, \end{cases} \tag{9.0.19}$$

with initial data (9.0.2), where $P(\rho)$ is given in (9.0.3).

We have the main result in this chapter:

Theorem 9.0.1 *Let the initial data $(\rho_0(x), u_0(x))$ be bounded measurable and $\rho_0(x) \geq 0$. Let $P(\rho) = \int_0^\rho s^2 e^s ds$ or $P(\rho) = \int_0^\rho s^2(s + d)^{\gamma-3}ds$, $\gamma > 3$. Then for fixed $\varepsilon > 0$, the unique smooth viscosity solution $(\rho^\varepsilon(x,t), u^\varepsilon(x,t))$ of the Cauchy problem (9.0.19), (9.0.2) exists and satisfies*

$$0 \leq \rho^\varepsilon(x,t) \leq M, \quad \|u^\varepsilon(x,t)\| \leq M, \tag{9.0.20}$$

where M is a positive constant, being independent of ε. Moreover, there exists a subsequence (still labelled) $(\rho^\varepsilon(x,t), u^\varepsilon(x,t))$ which converges almost everywhere on any bounded and open set $\Omega \subset R \times R^+$:

$$(\rho^\varepsilon(x,t), u^\varepsilon(x,t)) \to (\rho(x,t), u(x,t)), \quad as \ \varepsilon \downarrow 0^+, \tag{9.0.21}$$

where the limit pair of functions $(\rho(x,t), u(x,t))$ is a weak solution of the Cauchy problem (9.0.1), (9.0.2).

In Section 9.1, we shall first prove the L^∞ estimates (9.0.20) and hence the existence of viscosity solutions for the Cauchy problem (9.0.19), (9.0.2). From the constructions of entropy-entropy flux pair of Lax type obtained in Sections 9.2-9.3, we can see that all these entropy functions are regular in the range $\rho \geq 0$, although they are nonstrictly hyperbolic on the vacuum line $\rho = 0$. So the pointwise convergence of the viscosity solutions as ε tends to zero follows immediately from the same fashion as in Theorem 6.0.1 or Theorem 7.0.1.

9.1 Existence of Viscosity Solutions

In this section we consider the existence of viscosity solutions for the Cauchy problem (9.0.19), (9.0.2). By Theorem 1.0.2, it is sufficient to prove the L^∞ estimates given in (9.0.20).

When $P(\rho) = \int_0^\rho s^2 e^s ds$, we multiply system (9.0.19) by the vectors (w_ρ, w_u) and (z_ρ, z_u), respectively, where w, z are given by (9.0.8), to obtain

$$w_t + \lambda_2 w_x = \varepsilon w_{xx} - \frac{\varepsilon}{2} e^{\frac{\rho}{2}} \rho_x^2 \leq \varepsilon w_{xx} \qquad (9.1.1)$$

and

$$z_t + \lambda_1 z_x = \varepsilon z_{xx} + \frac{\varepsilon}{2} e^{\frac{\rho}{2}} \rho_x^2 \geq \varepsilon z_{xx}. \qquad (9.1.2)$$

Therefore applying the maximum principle to (9.1.1) and (9.1.2), respectively, we have $w(\rho^\varepsilon, u^\varepsilon) \leq M$ and $z(\rho^\varepsilon, u^\varepsilon) \geq -M$ for a suitable large constant depending only on L^∞ bound of the initial data. The nonnegativity of $\rho^\varepsilon \geq 0$ is obvious from $\rho_0(x) \geq 0$. Thus we have that

$$\Sigma_5 = \{(\rho, u) : w(\rho, u) \leq M, z(\rho, u) \geq -M, \rho \geq 0\}$$

is an invariant region (see Figure 9.1), where $w = u + 2e^{\frac{\rho}{2}}$, $z = u - 2e^{\frac{\rho}{2}}$ and hence, the boundedness of $(\rho^\varepsilon(x,t), u^\varepsilon(x,t))$.

When $P(\rho) = \int_0^\rho s^2(s+d)^{\gamma-3} ds$, $\gamma > 3$, exactly the same as the above case, we multiply system (9.0.19) by the vectors (w_ρ, w_u) and (z_ρ, z_u), respectively, where w, z are given by (9.0.16), to obtain

$$w_t + \lambda_2 w_x = \varepsilon w_{xx} - \frac{\varepsilon(\gamma-3)}{2}(\rho+d)^{\frac{\gamma-5}{2}} \rho_x^2 \leq \varepsilon w_{xx}, \qquad (9.1.3)$$

and

$$z_t + \lambda_1 z_x = \varepsilon z_{xx} + \frac{\varepsilon(\gamma-3)}{2}(\rho+d)^{\frac{\gamma-5}{2}} \rho_x^2 \geq \varepsilon z_{xx}. \qquad (9.1.4)$$

Therefore to apply the maximum principle to (9.1.3) and (9.1.4), respectively, we have that

$$D_1 = \{(\rho, u) : w(\rho, u) \leq M, z(\rho, u) \geq -M, \rho \geq 0\}$$

is an invariant region, which has a similar figure as Σ_5 in Figure 9.1, and hence complete the proof of (9.0.20). ■

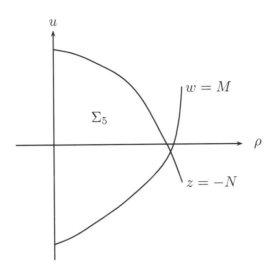

FIGURE 9.1

9.2   Lax Entropy for $P(\rho) = \int_0^\rho s^2 e^s ds$

In this section, we shall prove the second part of Theorem 9.0.1 for the case of $P(\rho) = \int_0^\rho s^2 e^s ds$, namely the pointwise convergence of the viscosity solutions $(\rho^\varepsilon(x,t), u^\varepsilon(x,t))$ as ε tends to zero. Exactly the same as the proof of Theorem 6.0.1, it is enough to construct four families of Lax entropy-entropy flux pairs and to prove the compactness of $\eta(\rho^\varepsilon(x,t), u^\varepsilon(x,t))_t + q(\rho^\varepsilon(x,t), u^\varepsilon(x,t))_x$ in $W_{loc}^{-1,2}(R \times R^+)$, for these entropy-entropy flux pairs, with respect to the sequence of viscosity solutions $(\rho^\varepsilon(x,t), u^\varepsilon(x,t))$.

A pair (η, q) of real-valued maps is an entropy-entropy flux pair of system (9.0.1) if

$$(q_\rho, q_u) = \left(u\eta_\rho + \frac{P'(\rho)}{\rho}\eta_u, \rho\eta_\rho + u\eta_u\right). \qquad (9.2.1)$$

If $P(\rho) = \int_0^\rho s^2 e^s ds$, the above system of equations is reduced to

$$(q_\rho, q_u) = \left(u\eta_\rho + \rho e^\rho \eta_u, \rho\eta_\rho + u\eta_u\right). \qquad (9.2.2)$$

Eliminating the q from (9.2.2), we have

$$\eta_{\rho\rho} = e^\rho \eta_{uu}. \tag{9.2.3}$$

If k denotes a positive constant, then the function $\eta = h(\rho)e^{ku}$ solves (9.2.3) provided that

$$h''(\rho) = k^2 e^\rho h. \tag{9.2.4}$$

Let $a(\rho) = e^{-\frac{1}{4}\rho}, r = 2ke^{\frac{1}{2}\rho}$ and $h(\rho) = a(\rho)\phi(r)$. Then ϕ satisfies a standard Fuchsian equation

$$\phi'' - (1 - \frac{1}{4r^2})\phi = 0. \tag{9.2.5}$$

Exactly the same as what we did in Chapters 6 and 7, we have a series solution of Equation (9.2.5) as follows:

$$\phi_1 = r^l(1 + \sum_{n=1}^\infty c_n r^n), \tag{9.2.6}$$

where

$$l = \frac{1}{2}, \quad c_2 = \frac{1}{2^2}, \quad c_{2n-1} = 0, \quad c_{2n} = \frac{c_{2(n-1)}}{(2n)^2} = \frac{1}{2^{2n}(n!)^2} \text{ for } n \geq 2. \tag{9.2.7}$$

Let another independent solution ϕ_2 of (9.2.5) satisfy $\phi_2 = \phi_1 Q$. Then Q solves

$$Q'' + \frac{2\phi_1'}{\phi_1}Q' = 0. \tag{9.2.8}$$

Thus

$$Q' = -(\phi_1)^{-2} = -(rg^2(r))^{-1}, \tag{9.2.9}$$

where

$$g(r) = 1 + \sum_{n=1}^\infty c_{2n} r^{2n}. \tag{9.2.10}$$

Let

$$Q = \int_r^\infty (rg^2(r))^{-1} dr. \tag{9.2.11}$$

Then

$$\phi_2 = r^{\frac{1}{2}} g(r) \int_r^\infty (rg^2(r))^{-1} dr. \tag{9.2.12}$$

Using (9.2.2), we have

$$q_u = \rho \eta_\rho + u \eta_u \tag{9.2.13}$$

and two progressing waves of system (9.0.1) are provided by

$$\begin{cases} \eta_k = h(\rho) e^{ku} \\ q_k = u\eta_k + (\rho h' - h) e^{ku}/k \end{cases} \tag{9.2.14}$$

and

$$\begin{cases} \eta_{-k} = h(\rho) e^{-ku} \\ q_{-k} = u\eta_{-k} + (h - \rho h') e^{-ku}/k. \end{cases} \tag{9.2.15}$$

Since $h(\rho) = a(\rho)\phi(r), a(\rho) = e^{-\frac{1}{4}\rho}$ and $r = 2ke^{\frac{1}{2}\rho}$, then

$$q_k = (u + \rho e^{\frac{\rho}{2}} \frac{\phi'(r)}{\phi(r)} - \frac{4+\rho}{4k})\eta_k \tag{9.2.16}$$

and

$$q_{-k} = (u - \rho e^{\frac{\rho}{2}} \frac{\phi'(r)}{\phi(r)} + \frac{4+\rho}{4k})\eta_{-k}. \tag{9.2.17}$$

Let $\eta_k^1 = a(\rho)\phi_1(r)e^{ku}$. Then using Lemma 6.2.1, we have

$$\eta_k^1 = a(\rho)\phi_1(r)e^{-r}e^{kw} = e^{kw}(a(\rho) + O(\frac{1}{k})) \tag{9.2.18}$$

on $\rho \geq 0$ as k approaches infinity and the corresponding flux function is of the form

$$\begin{aligned} q_k^1 &= \eta_k^1 (\lambda_2 + \rho e^{\frac{\rho}{2}} (\frac{\phi_1'(r)}{\phi_1(r)} - 1) - \frac{4+\rho}{4k}) \\ &= \eta_k^1 (\lambda_2 - \frac{4+\rho}{4k} + O(\frac{1}{k^2})) \end{aligned} \tag{9.2.19}$$

on $\rho \geq 0$. Similarly let $\eta_{-k}^1 = a(\rho)\phi_1(r)e^{-ku}$. Then

$$\eta_{-k}^1 = a(\rho)\phi_1(r)e^{-r}e^{-kz} = e^{-kz}(a(\rho) + O(\frac{1}{k})) \tag{9.2.20}$$

on $\rho \geq 0$ and the corresponding flux function

$$
\begin{aligned}
q^1_{-k} &= \eta^1_{-k}(\lambda_1 - \rho e^{\frac{\rho}{2}}(\frac{\phi'_1(r)}{\phi_1(r)} - 1) + \frac{4+\rho}{4k}) \\
&= \eta^1_{-k}(\lambda_1 + \frac{4+\rho}{4k} + O(\frac{1}{k^2}))
\end{aligned}
\tag{9.2.21}
$$

on $\rho \geq 0$. The entropy-entropy flux pairs $(\eta^2_k, q^2_k), (\eta^2_{-k}, q^2_{-k})$ satisfy

$$
\eta^2_k = a(\rho)\phi_2(r)e^{ku} = e^{kz}(a(\rho) + O(\frac{1}{k}))
\tag{9.2.22}
$$

on $\rho \geq 0$;

$$
q^2_k = \eta^2_k(\lambda_1 - \frac{4+\rho}{4k} + O(\frac{1}{k^2}))
\tag{9.2.23}
$$

on $\rho \geq 0$;

$$
\eta^2_{-k} = a(\rho)\phi_2(r)e^{-ku} = e^{-kw}(a(\rho) + O(\frac{1}{k}))
\tag{9.2.24}
$$

on $\rho \geq 0$;

$$
q^2_{-k} = \eta^2_{-k}(\lambda_2 + \frac{4+\rho}{4k} + O(\frac{1}{k^2}))
\tag{9.2.25}
$$

on $\rho \geq 0$ respectively.

Furthermore, we have from (9.2.18)-(9.2.25) that

$$
\begin{cases}
q^1_k = \lambda_2\eta^1_k - e^{kw}((\frac{4+\rho}{4k})e^{-\frac{1}{4}\rho}/k + O(\frac{1}{k^2})), \\
q^1_{-k} = \lambda_1\eta^1_{-k} + e^{-kz}((\frac{4+\rho}{4k})e^{\frac{1}{4}\rho}/k + O(\frac{1}{k^2})), \\
q^2_k = \lambda_1\eta^2_k - e^{kz}((\frac{4+\rho}{4k})e^{-\frac{1}{4}\rho}/k + O(\frac{1}{k^2})), \\
q^2_{-k} = \lambda_2\eta^2_{-k} + e^{-kw}((\frac{4+\rho}{4k})e^{-\frac{1}{4}\rho}/k + O(\frac{1}{k^2}))
\end{cases}
\tag{9.2.26}
$$

on $\rho \geq 0$.

It is easy to check that system (9.0.1) for the case $P(\rho) = \int_0^\rho s^2 e^s ds$ has a strictly convex entropy

$$
\eta^\star = \frac{1}{2}u^2 + e^\rho
$$

and the corresponding entropy flux

$$q^\star = \frac{1}{3}u^3 + \rho u e^\rho.$$

From this strictly entropy-entropy flux pair and the method given in Chapters 6 and 7, we deduce that

$$\varepsilon^{\frac{1}{2}}\partial_x \rho^\varepsilon, \varepsilon^{\frac{1}{2}}\partial_x u^\varepsilon \quad \text{are uniformly bounded in } L^2_{loc}(R \times R^+). \quad (9.2.27)$$

Noticing that all entropy-entropy flux pairs constructed above are smooth in the range $\rho \geq 0$, we have the following lemma, which combining with the compensated compactness method given in Chapters 6 and 7 completes the proof of Theorem 9.0.1 for the case $P(\rho) = \int_0^\rho s^2 e^s ds$.

Lemma 9.2.1 *For any entropy-entropy flux pair $(\eta(\rho, u), q(\rho, u))$ given in (9.2.18)-(9.2.25),*

$$\eta(\rho^\varepsilon, u^\varepsilon)_t + q(\rho^\varepsilon, u^\varepsilon)_x \text{ is compact in } H^{-1}_{loc}(R \times R^+) \quad (9.2.28)$$

with respect to the sequence of viscosity solutions $(\rho^\varepsilon, u^\varepsilon)$.

9.3   Lax Entropy for $P(\rho) = \int_0^\rho s^2(s+d)^{\gamma-3}ds$

In this section, we construct four families of the entropy-entropy flux pair of Lax type and prove the compactness of $\eta(\rho^\varepsilon(x,t), u^\varepsilon(x,t))_t + q(\rho^\varepsilon(x,t), u^\varepsilon(x,t))_x$ in $W^{-1,2}_{loc}(R \times R^+)$ for the case of $P(\rho) = \int_0^\rho s^2(s+d)^{\gamma-3}ds$ in system (9.0.1).

If $P(\rho) = \int_0^\rho s^2(s+d)^{\gamma-3}ds$, $\gamma > 3$, system (9.2.1) is reduced to

$$(q_\rho, q_u) = (u\eta_\rho + \rho(\rho+d)^{\gamma-3}\eta_u, \rho\eta_\rho + u\eta_u). \quad (9.3.1)$$

Eliminating the q from (9.3.1), we have

$$\eta_{\rho\rho} = (\rho+d)^{\gamma-3}\eta_{uu}. \quad (9.3.2)$$

If k denotes a positive constant, then the function $\eta = h(\rho)e^{ku}$ solves (9.3.2) provided that

$$h''(\rho) = k^2(\rho+d)^{\gamma-3}h. \quad (9.3.3)$$

Let $a(\rho) = (\rho + d)^{\frac{1}{4}(3-\gamma)}$, $s = \frac{2k}{\gamma-1}(\rho + d)^{\frac{1}{2}(\gamma-1)}$. Then $h = a(\rho)\phi(s)$ solves (9.3.3) if and only if ϕ satisfies the standard Fuchsian equation

$$\phi'' - (1 + \mu s^2)\phi = 0, \tag{9.3.4}$$

where $\mu = \frac{4-(\gamma-1)^2}{4(\gamma-1)^2} > -\frac{1}{4}$.

From (9.3.1), we have

$$q_u = \rho\eta_\rho + u\eta_u. \tag{9.3.5}$$

If

$$\eta_k = h(\rho)e^{ku}, \tag{9.3.6}$$

then

$$(q_k)_u = \rho h'(\rho)e^{ku} + kuh(\rho)e^{ku}$$

and hence one entropy flux corresponding to η_k is

$$q_k = uh(\rho)e^{ku} + (\rho h' - h)e^{ku}/k$$

$$= \eta_k\left(u + \rho(\rho+d)^{\frac{1}{2}(\gamma-3)}\frac{\phi'}{\phi} - \frac{(\gamma+1)\rho + 4d}{4k(\rho+d)}\right). \tag{9.3.7}$$

If

$$\eta_{-k} = h(\rho)e^{-ku}, \tag{9.3.8}$$

then

$$(q_{-k})_u = \rho h'(\rho)e^{-ku} - kuh(\rho)e^{-ku}$$

and hence one entropy flux corresponding to η_{-k} is

$$q_{-k} = uh(\rho)e^{-ku} + (h - \rho h')e^{-ku}/k$$

$$= \eta_{-k}\left(u - \rho(\rho+d)^{\frac{1}{2}(\gamma-3)}\frac{\phi'}{\phi} + \frac{(\gamma+1)\rho + 4d}{4k(\rho+d)}\right). \tag{9.3.9}$$

Then two progressing waves of system (9.0.1) for the case of $P(\rho) = \int_0^\rho s^2(s+d)^{\gamma-3}ds$ are provided by (9.3.6)-(9.3.7) and (9.3.8)-(9.3.9).

We may use the method of Frobenius again to obtain a series solution of Equation (9.3.4) as follows:

$$\phi_1 = s^j \sum_{n=0}^{\infty} e_n s^n, \tag{9.3.10}$$

where e_0 can be any positive constant, $j > 0$ is any one root of the equation $j(j-1) = \mu$, and e_n satisfy

$$e_n = \frac{e_{n-1}}{(2n+j)(2n+j-1)-\mu} \quad \text{for } n \geq 1. \tag{9.3.11}$$

Let another independent solution ϕ_2 of (9.3.4) satisfy $\phi_2 = \phi_1 Q$. Then Q solves

$$Q'' + \frac{2\phi_1'}{\phi_1}Q' = 0. \tag{9.3.12}$$

Thus

$$Q' = -(\phi_1)^{-2} = -(s^j g(s))^{-2}, \tag{9.3.13}$$

where

$$g(s) = \sum_{n=0}^{\infty} e_n s^n. \tag{9.3.14}$$

Let

$$Q = \int_s^{\infty} (s^j g(s))^{-2} ds. \tag{9.3.15}$$

Then

$$\phi_2 = s^j g(s) \int_s^{\infty} (s^j g(s))^{-2} ds. \tag{9.3.16}$$

It is easy to check that $\phi_1(s) > 0, \phi_1'(s) > 0, \phi_2(s) > 0$ and $\phi_2'(s) < 0$.

Let $\eta_k^1 = a(\rho)\phi_1(s)e^{ku}$. Then using Lemma 6.2.1, we have

$$\eta_k^1 = a(\rho)\phi_1(s)e^{-s}e^{kw} = e^{kw}(a(\rho) + O(\frac{1}{k})) \tag{9.3.17}$$

on $\rho \geq 0$ as k approaches infinity and the corresponding flux function is of the form

$$q_k^1 = \eta_k^1\left(u + \rho(\rho+d)^{\frac{1}{2}(\gamma-3)}\right.$$

$$+ \rho(\rho+d)^{\frac{1}{2}(\gamma-3)}(\frac{\phi_1'(s)}{\phi_1(s)} - 1) - \left.\frac{(\gamma+1)\rho+4d}{4k(\rho+d)}\right) \tag{9.3.18}$$

$$= \eta_k^1(\lambda_2 - \frac{(\gamma+1)\rho+4d}{4k(\rho+d)} + O(\frac{1}{k^2}))$$

on $\rho \geq 0$. Similarly let $\eta^1_{-k} a(\rho)\phi_1(s)e^{-ku}$. Then

$$\eta^1_{-k} = a(\rho)\phi_1(s)e^{-s}e^{-kz} = e^{-kz}(a(\rho) + O(\frac{1}{k})) \tag{9.3.19}$$

on $\rho \geq 0$ and the corresponding flux function

$$q^1_{-k} = \eta^1_{-k}\left(u - \rho(\rho+d)^{\frac{1}{2}(\gamma-3)}\right.$$

$$-\rho(\rho+d)^{\frac{1}{2}(\gamma-3)}\left(\frac{\phi_1'(s)}{\phi_1(s)} - 1\right) + \frac{(\gamma+1)\rho + 4d}{4k(\rho+d)}\right) \tag{9.3.20}$$

$$= \eta^1_{-k}(\lambda_1 + \frac{(\gamma+1)\rho + 4d}{4k(\rho+d)} + O(\frac{1}{k^2}))$$

on $\rho \geq 0$. The entropy-entropy flux pairs $(\eta^2_k, q^2_k), (\eta^2_{-k}, q^2_{-k})$ satisfy

$$\eta^2_k = a(\rho)\phi_2(s)e^u = a(\rho)\phi_2(s)e^s e^{kz} = e^{kz}(a(\rho) + O(\frac{1}{k})) \tag{9.3.21}$$

on $\rho \geq 0$;

$$q^2_k = \eta^2_k(\lambda_1 - \frac{(\gamma+1)\rho + 4d}{4k(\rho+d)} + O(\frac{1}{k^2})) \tag{9.3.22}$$

on $\rho \geq 0$;

$$\eta^2_{-k} = a(\rho)\phi_2(s)e^{-u} = e^{-kw}(a(\rho) + O(\frac{1}{k})) \tag{9.3.23}$$

on $\rho \geq 0$;

$$q^2_{-k} = \eta^2_{-k}(\lambda_2 + \frac{(\gamma+1)\rho + 4d}{4k(\rho+d)} + O(\frac{1}{k^2})) \tag{9.3.24}$$

on $\rho \geq 0$ respectively.

It is easy to check that system (9.0.1) for the case of $P(\rho) = \int_0^\rho s^2(s+d)^{\gamma-3}ds$ has a strictly convex entropy

$$\eta^\star = \frac{1}{(\gamma-2)(\gamma-1)}(\rho+d)^{\gamma-1} + \frac{1}{2}u^2.$$

Then we have the following lemma:

Lemma 9.3.1

$\varepsilon^{\frac{1}{2}}\partial_x u^\varepsilon$, $\varepsilon^{\frac{1}{2}}\partial_x \rho^\varepsilon$ *are uniformly bounded in* $L^2_{loc}(R \times (0, \infty))$.   (9.3.25)

Noticing that all entropy-entropy flux pairs constructed above are smooth in the range $\rho \geq 0$, we have the following lemma, which combining with the compensated compactness method given in Chapters 6 and 7 completes the proof of Theorem 9.0.1 for the case $P(\rho) = \int_0^\rho s^2(s+d)^{\gamma-3}ds$, $\gamma > 3$.

Lemma 9.3.2 *For any entropy-entropy flux pair* $(\eta(\rho, u), q(\rho, u))$ *given in (9.3.17)-(9.3.24),*

$$\eta(\rho^\varepsilon, u^\varepsilon)_t + q(\rho^\varepsilon, u^\varepsilon)_x \text{ is compact in } H^{-1}_{loc}(R \times R^+).   \quad (9.3.26)$$

9.4   Related Results

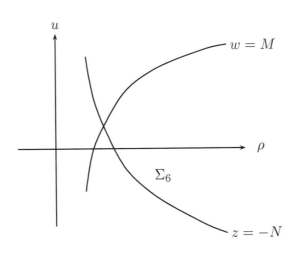

FIGURE 9.2

The large data existence theorem of global weak solutions with locally finite total variation for the Cauchy problem (9.0.1), (9.0.2) was established by DiPerna [Di1] for general pressure function $P(\rho)$

satisfying suitable conditions, for instance, $P(\rho) = k^2 \rho^\gamma, \gamma \in (1,3)$, such that the invariant region

$$\Sigma_6 = \{(\rho, u) : w(\rho, u) \le M, z(\rho, u) \ge M\} \tag{9.4.1}$$

is away from the vacuum line $\rho = 0$ as given in Figure 9.2.

For this class of functions $P(\rho)$, the Glimm method works well since system (9.0.1) is strictly hyperbolic and two eigenvalues are distant in the range Σ_6. The uniqueness of the weak solution for this case was recently obtained by Bressan [Br].

However, for the case of $\gamma > 3$ or more general pressure function $P(\rho)$ given in the next chapter, the Glimm method does not work because the invariant region always includes the nonstrictly hyperbolic line $\rho = 0$ as graphed in Figure 9.1 for $P(\rho) = \int_0^\rho s^2 e^s ds$ or $P(\rho) = \int_0^\rho s^2 (s+d)^{\gamma-3} ds$.

The proofs in this chapter are from [Lu8] for the case $P(\rho) = \int_0^\rho s^2 e^s ds$ and from [Lu2] for the case of $P(\rho) = \int_0^\rho s^2 (s+d)^{\gamma-3} ds$.

Chapter 10

General Euler Equations of Compressible Fluid Flow

In this chapter, we consider the existence of global weak solutions for the following nonlinear hyperbolic systems:

$$\begin{cases} u_t + (\frac{1}{2}u^2 + f(v))_x = 0 \\ v_t + (uv + g(v))_x = 0, \end{cases} \tag{10.0.1}$$

with bounded measurable initial data

$$(u(x,0), v(x,0)) = (u_0(x), v_0(x)). \tag{10.0.2}$$

When $g(v) = 0$, system (10.0.1) is the same as system (9.0.1) with v in (10.0.1) instead of ρ in (9.0.1).

In Chapters 6, 7 and 9, we apply the compensated compactness method to the system of quadratic flux, Le Roux system and two special systems of Euler equations. From these applications, we can see that because of the explicit constructions of flux functions in these systems, we can make some suitable transformations of variables to construct explicit entropy-entropy flux pairs of Lax type via the solutions of Fuchsian equations, and hence obtain necessary estimates about these function pairs. Then based on these estimates and using some developed ideas from the theory of compensated compactness, we can prove the large data existence theorem of global weak solutions for these systems. In Chapter 8, the explicit flux function $P(\rho) = k^2\rho^\gamma$ in the system of gas dynamics also helps much to obtain the explicit expression of weak entropies, which is clearly the crux to establish the compactness of the sequence of viscosity solutions and hence the existence of weak solutions.

However, system (10.0.1) is in a different situation from systems we considered in the above chapters, in which the flux functions are nonlinear and in implicit forms.

Entropy-entropy flux pairs to more general strictly hyperbolic systems or systems in the strictly hyperbolic domains were well analyzed by Lax ([La2, La3]). However, to apply the compensated compactness method to some nonstrictly hyperbolic systems just as given in the form of system (10.0.1), some new techniques to construct entropy-entropy flux pairs must be investigated.

One new idea in this chapter is to find entropy-entropy flux pairs to system (10.0.1) in the following special form:

$$\eta_k^1 = e^{kw}\left(a_1(v) + \frac{b_1(v,k)}{k}\right), \quad q_k^1 = e^{kw}\left(c_1(v) + \frac{d_1(v,k)}{k}\right);$$

$$\eta_{-k}^2 = e^{-kw}\left(a_2(v) + \frac{b_2(v,k)}{k}\right), \quad q_{-k}^2 = e^{-kw}\left(c_2(v) + \frac{d_2(v,k)}{k}\right);$$

$$\eta_{-k}^1 = e^{-kz}\left(a_3(v) + \frac{b_3(v,k)}{k}\right), \quad q_{-k}^1 = e^{-kz}\left(c_3(v) + \frac{d_3(v,k)}{k}\right);$$

$$\eta_k^2 = e^{kz}\left(a_4(v) + \frac{b_4(v,k)}{k}\right), \quad q_k^2 = e^{kz}\left(c_4(v) + \frac{d_4(v,k)}{k}\right),$$

where w, z are the Riemann invariants of system (10.0.1) given by (10.1.1). Notice that all the unknown functions $a_i, b_i (i = 1, 2, 3, 4)$ are only of a single variable v. This special simple construction yields an ordinary differential equation of second order with a singular coefficient $1/k$ before the term of the second order derivative. Then the necessary estimates for functions $a_i(v), b_i(v,k)$ are obtained by the use of the singular perturbation theory of ordinary differential equations.

For system (10.0.1), our assumptions about $g(v), f(v)$ and initial data are as follows:

(A_1) $f, g \in C^3[0, \infty)$ and $f_1 = (f'/v) \in C^2[0, \infty), g_1 = (g'/v) \in C^2[0, \infty)$, satisfy $f_1 \geq d$ and

$$2f_1' + g_1'(s_1 + g_1) \geq 0, \text{ for } v \geq 0,$$

$$2f_1' + g_1'(s_1 - g_1) \geq 0, \text{ for } v \geq 0,$$

where d is a fixed positive constant, and $s_1 = \sqrt{g_1^2 + 4f_1}$.

(A_2) u_0, v_0 are bounded measurable and $v_0 \geq 0$.

Example 10.0.1 *Besides the general Euler equations of compressible fluid flow (9.0.1), there are many other function pairs (f, g) which satisfy the condition (A_1). For example, we can specially choose $f_1 = (v + d)^l, g_1 = k(v + e)^m$, where d, e, m, l are positive constants, k is a non-negative constant and $e \geq d, l \geq m$. Then it is easy to check that (A_1) is satisfied.*

By simple calculations, two eigenvalues of system (10.0.1) are

$$\lambda_1 = \frac{2u + vg_1(v) - vs_1}{2}, \quad \lambda_2 = \frac{2u + vg_1(v) + vs_1}{2} \qquad (10.0.3)$$

with corresponding right eigenvectors

$$r_1 = (-2f_1, s_1 - g_1)^T, \quad r_2 = (2f_1, s_1 + g_1)^T. \qquad (10.0.4)$$

So, using the assumption (A_1) we have

$$\begin{aligned}
\nabla \lambda_1 \cdot r_1 &= -2f_1 + (s_1 - g_1)[\frac{vg' + g_1 - s_1}{2} - \frac{v(g_1 g_1' + 2f_1')}{2s_1}] \\
&= \frac{1}{2s_1} - 4f_1 s_1 + (s_1 - g_1)[s_1(vg_1' + g_1) - s_1^2 - vg_1 g_1' - 2vf_1'] \\
&\leq \frac{1}{2s_1}[-4f_1 s_1 + (s_1 - g_1)(s_1 g_1 - s_1^2)] \\
&= \frac{1}{2}(-4f_1 - (s_1 - g_1)^2) < 0
\end{aligned}$$

$$(10.0.5)$$

and

$$\begin{aligned}
\nabla \lambda_2 \cdot r_2 &= 2f_1 + (s_1 + g_1)[\frac{vg' + g_1 + s_1}{2} + \frac{v(g_1 g_1' + 2f_1')}{2s_1}] \\
&\geq \frac{1}{2}(4f_1 + (s_1 + g_1)^2) > 0.
\end{aligned}$$

$$(10.0.6)$$

Therefore, it follows from (10.0.3) that $\lambda_1 = \lambda_2$ at the line $v = 0$, in which the strict hyperbolicity for system (10.0.1) fails to hold. However, both characteristic fields are genuinely nonlinear from (10.0.5) and (10.0.6).

We add small, positive perturbation terms to system (10.0.1) and consider the Cauchy problem for the following system:

$$\begin{cases} u_t + (\frac{1}{2}u^2 + f(v))_x = \varepsilon u_{xx}, \\ v_t + (uv + g(v))_x = \varepsilon v_{xx}, \end{cases} \quad (10.0.7)$$

with initial data (10.0.2).

Then we have the main result in this chapter as follows:

Theorem 10.0.2 *Let the assumptions $(A_1), (A_2)$ hold. Then, for any fixed $\varepsilon > 0$, the Cauchy problem (10.0.7), (10.0.2) has a unique global smooth solution $(u^\varepsilon(x,t), v^\varepsilon(x,t))$ satisfying*

$$|u^\varepsilon(x,t)| \le M, \quad 0 \le v^\varepsilon(x,t) \le M, \quad (10.0.8)$$

where M is a positive constant, independent of ε.

Moreover, there exists a subsequence (still labelled) $(u^\varepsilon(x,t), v^\varepsilon(x,t))$ such that

$$(u^\varepsilon(x,t), v^\varepsilon(x,t)) \to (u(x,t), v(x,t)), \quad a.e. \ on \ \Omega, \quad (10.0.9)$$

where $\Omega \subset R \times R^+$ is any bounded and open set, the limit pair of functions $(u(x,t), v(x,t))$ being a weak solution of the Cauchy problem (10.0.1), (10.0.2).

The existence of viscosity solutions for the Cauchy problem (10.0.7), (10.0.2) is given in Section 10.1. In Section 10.2, we shall construct entropy-entropy flux pairs of system (10.0.1) and obtain necessary estimates. In Section 10.3, these estimates will yield the existence of weak solutions for the Cauchy problem (10.0.1)-(10.0.2) when coupled with the method to deal with nonstrictly hyperbolic systems given in Chapters 6, 7 and 9.

10.1 Existence of Viscosity Solutions

In this section, we shall prove the first part of Theorem 10.0.2, namely the existence of smooth viscosity solutions $(u^\varepsilon, v^\varepsilon)$ for the Cauchy problem (10.0.7), (10.0.2).

By Theorem 1.0.2, the unique thing is to obtain the *a priori* L^∞ estimate (10.0.8).

By simple calculations, two Riemann invariants of system (10.0.1) are

$$w = u + \int_0^v \frac{g_1 + s_1}{2} dv, \quad z = u + \int_0^v \frac{g_1 - s_1}{2} dv. \quad (10.1.1)$$

Along the line of $w = M$, we have from assumption (A_1) that

$$\frac{du}{dv} = -\frac{g_1 + s_1}{2} \le 0, \quad \frac{d^2u}{dv^2} = -\frac{2f_1' + g_1'(g_1 + s_1)}{2s_1} \le 0, \text{ as } v \ge 0,$$

and along the line of $z = -M$, we have from assumption (A_1) that

$$\frac{du}{dv} = \frac{-g_1 + s_1}{2} \le 0, \quad \frac{d^2u}{dv^2} = \frac{2f_1' + g_1'(g_1 - s_1)}{2s_1} \ge 0, \text{ as } v \ge 0.$$

So by the theory of invariant regions introduced in Theorem 4.2.1, it is easy to get that

$$D_2 = \{(u, v) : w(u, v) \le M, \quad z(u, v) \ge -M, \quad v \ge 0\}$$

is an invariant region, which has a similar figure as D_1 or Σ_5 given in Figure 9.1, where M is a suitable large positive constant.

This invariant region D_2 yields the L^∞ bound (10.0.8) and hence the proof of existence of viscosity solutions.

10.2 Lax Entropy and Related Estimates

To prove the existence of weak solutions in the second part of Theorem 10.0.2, in this section, we shall construct the entropy entropy flux pairs of Lax type for system (10.0.1) and give the required estimates by means of the theory of singular perturbation.

We recall that a pair (η, q) of real-valued maps is an entropy-entropy flux pair of (10.0.1) if all smooth solutions satisfy

$$(u\eta_u + v\eta_v, f'\eta_u + (u + g')\eta_v) = (q_u, q_v). \quad (10.2.1)$$

Eliminating the q from (10.2.1), we have

$$f'\eta_{uu} + g'\eta_{uv} - v\eta_{vv} = 0. \quad (10.2.2)$$

Substituting entropies $\eta_k^1 = e^{kw}(a_1(v) + b_1(v, k)/k)$ into (10.2.2), we obtain that

$$k[s_1 a_1' + \frac{g_1'(g_1 + s_1) + 2f_1'}{2s_1} a_1]$$

$$+ a_1'' + s_1 b_1' + \frac{g_1'(g_1 + s_1) + 2f_1'}{2s_1} b_1 + \frac{b_1''}{k} \qquad (10.2.3)$$

$$= 0.$$

Let

$$s_1 a_1' + \frac{g_1'(g_1 + s_1) + 2f_1'}{2s_1} a_1 = 0 \qquad (10.2.4)$$

and

$$a_1'' + s_1 b_1' + \frac{g_1'(g_1 + s_1) + 2f_1'}{2s_1} b_1 + \frac{b_1''}{k} = 0. \qquad (10.2.5)$$

Then

$$a_1 = \exp(-\int_0^v \frac{g_1'(g_1 + s_1) + 2f_1'}{2s_1^2} dv) > 0 \text{ for } v \geq 0. \qquad (10.2.6)$$

The existence of b_1 and its uniform bound with respect to k can be obtained by the following lemma (cf. [Ka]):

Lemma 10.2.1 *Let $Y(x) \in C^2[0, h]$ be the solution of the equation*

$$F(x, Y, Y') = 0,$$

and functions $f(x, y, z, \lambda), F(x, y, z)$ be continuous on the regions $0 \leq x \leq h, |y - Y(x)| \leq l(x), |z - Y'(x)| \leq m(x)$ for some positive functions $l(x), m(x)$ and $\lambda_0 > \lambda > 0$. In addition,

$$|f(x, y, z, \lambda) - F(x, y, z)| \leq \varepsilon,$$

$$|F(x, y_2, z) - F(x, y_1, z)| \leq M|y_2 - y_1|,$$

$$\frac{F(x, y, z_2) - F(x, y, z_1)}{z_2 - z_1} \geq L$$

for some positive constants ε, M and L.

If $y(x) = y(x, \lambda)$ *is a solution of the following ordinary differential equation of second order:*

$$\lambda y'' + f(x, y, y', \lambda) = 0,$$

with $y(0) = Y(0)$ *and* $y'(0)$ *being arbitrary, then for sufficiently small* $\lambda > 0, \varepsilon > 0$ *and* $P = |y'(0) - Y'(0)|$, $y(x)$ *exists for all* $0 \leq x \leq h$ *and satisfies*

$$|y(x, \lambda) - Y(x)| < \left[\frac{\varepsilon}{M} + \lambda(\frac{P}{L} + \frac{N}{M})\right] exp(\frac{Mx}{L}),$$

where $N = \max\limits_{0 \leq x \leq h} |Y(x)|$.

Furthermore, we can use Lemma 10.2.1 again to obtain the bound of b_1' with respect to k if we differentiate Equation (10.2.5) with respect to v.

Using (10.2.1), we have

$$q_u = u\eta_u + v\eta_v, \quad q_v = f'\eta_u + (u + g')\eta_v.$$

Then a progressing wave of system (10.0.1) is provided by

$$\eta_k^1 = e^{kw}(a_1(v) + \frac{b_1(v, k)}{k}), \quad q_k^1 = \lambda_2\eta_k^1 + e^{kw}(\frac{va_1' - a_1}{k} + \frac{vb_1' - b_1}{k^2}). \tag{10.2.7}$$

In a similar way, we can obtain another entropy-entropy flux pair of Lax type as follows:

$$\eta_{-k}^2 = e^{-kw}(a_2(v) + \frac{b_2(v, k)}{k}),$$
$$q_{-k}^2 = \lambda_2\eta_{-k}^2 + e^{-kw}(\frac{a_2 - va_2'}{k} + \frac{b_2 - vb_2'}{k^2}), \tag{10.2.8}$$

where $a_2(v) = a_1(v)$ and $b_2(v, k)$ satisfies

$$a_1'' - s_1b_2' - \frac{g_1'(g_1 + s_1) + 2f_1'}{2s_1}b_2 + \frac{b_2''}{k} = 0; \tag{10.2.9}$$

$$\eta_k^2 = e^{kz}(a_3(v) + \frac{b_3(v, k)}{k}),$$
$$q_k^2 = \lambda_1\eta_k^2 + e^{kz}(\frac{va_3' - a_3}{k} + \frac{vb_3' - b_3}{k^2}), \tag{10.2.10}$$

where

$$s_1 a_3' + \frac{g_1'(g_1 - s_1) + 2f_1'}{2s_1} a_3 = 0 \tag{10.2.11}$$

and

$$a_3'' - s_1 b_3' - \frac{g_1'(g_1 - s_1) + 2f_1'}{2s_1} b_3 + \frac{b_3''}{k} = 0; \tag{10.2.12}$$

$$\eta_{-k}^1 = e^{-kz}\left(a_4(v) + \frac{b_4(v, k)}{k}\right),$$
$$q_{-k}^1 = \lambda_1 \eta_{-k}^1 + e^{-kz}\left(\frac{a_4 - va_4'}{k} + \frac{b_4 - vb_4'}{k^2}\right), \tag{10.2.13}$$

where $a_4(v) = a_3(v)$ and $b_4(v, k)$ satisfies

$$a_3'' + s_1 b_4' + \frac{g_1'(g_1 - s_1) + 2f_1'}{2s_1} b_4 + \frac{b_4''}{k} = 0. \tag{10.2.14}$$

From (10.2.11), we have

$$a_4 = a_3 = \exp\left(-\int_0^v \frac{g_1'(g_1 - s_1) + 2f_1'}{2s_1^2} dv\right) > 0 \text{ for } v \geq 0. \tag{10.2.15}$$

Using the argument in Lemma 10.2.1 in Equation (10.2.14), we can get the existence of b_4 and the uniform bounded estimates of b_4, b_4' with respect to k. If making an independent transformation $v_1 = v - M$ to Equations (10.2.9) and (10.2.12), where M is the upper bound of v, we also obtain the existence of b_2, b_3 and the uniform bounded estimates of b_2, b_3, b_2' and b_3' by Lemma 10.2.1 again.

Noticing the assumptions $(A_1), (A_2)$, we have that $a_i - va_i', i = 1, 2, 3, 4$ are all positive for $v \geq 0$.

10.3 Existence of Weak Solutions

In Section 10.2, four families of entropy-entropy flux pairs of Lax type for system (10.0.1) are constructed. Noticing the estimates about functions $a_i(v), a_i - va_i', i = 1, 2, 3, 4$, and the forms of these function pairs given by (10.2.7), (10.2.8), (10.2.10) and (10.2.13), we can use the same method given in Chapters 6, 7 and 9 to deduce that all the Young measures ν determined by the sequence of viscosity solutions of the Cauchy problem (10.0.7), (10.0.2) are Dirac measures if the following lemma is true.

Lemma 10.3.1 *For any entropy-entropy flux pair $(\eta(u,v), q(u,v))$ of system (10.0.1) given by (10.2.7), (10.2.8), (10.2.10) and (10.2.13),*

$$\eta(u^\varepsilon, v^\varepsilon)_t + q(u^\varepsilon, v^\varepsilon)_x \text{ is compact in } H_{loc}^{-1}(R \times R^+) \qquad (10.3.1)$$

with respect to the viscosity solutions $(u^\varepsilon, v^\varepsilon)$ of the Cauchy problem (10.0.7), (10.0.2).

Proof. It is easy to check that system (10.0.1) has a strictly convex entropy

$$\eta^\star = \frac{1}{2}u^2 + \int_0^v \int_0^y \frac{f'(s)}{s} ds dy. \qquad (10.3.2)$$

Then using this convex entropy, we can prove that

$$\varepsilon^{\frac{1}{2}}\partial_x u^\varepsilon, \ \varepsilon^{\frac{1}{2}}\partial_x v^\varepsilon \text{ are uniformly bounded in } L_{loc}^2(R \times R^+). \quad (10.3.3)$$

Notice that all entropy-entropy flux pairs given in (10.2.7), (10.2.8), (10.2.10) and (10.2.13) are smooth in the range $v \geq 0$. Thus we complete the proof of Lemma 10.2.2 by using Theorem 2.3.2.

10.4 Related Results

The proof in this chapter is from the paper [Lu6]. The ideas to construct entropy-entropy flux pairs to nonstrictly hyperbolic systems by means of the singular perturbation theory of ordinary differential equations of second order were extended in [KL2] to study some hyperbolic systems with inhomogeneous terms.

Chapter 11

Extended Systems of Elasticity

In this chapter, we consider the existence of global weak solutions for the following extended nonlinear hyperbolic systems of elasticity:

$$\begin{cases} u_t + (cu + f(v))_x = 0 \\ v_t + (u + g(v))_x = 0, \end{cases} \qquad (11.0.1)$$

with bounded measurable initial data

$$(u(x,0), v(x,0)) = (u_0(x), v_0(x)), \qquad (11.0.2)$$

where c is a constant.

When $g(v) = 0$ and $c = 0$, (11.0.1) is the system of one-dimensional nonlinear elasticity in Lagrangian coordinates which describes the balance of mass and linear momentum, where v denotes the strain, $f(v)$ is the stress and u the velocity.

For the more general system (11.0.1), our assumptions about $g(v)$ and $f(v)$ are as follows:

(A) $f, g \in C^3$ satisfy $f' \geq d$ for $v \in R$, and

$$2f'' + g''(s + g' - c) > 0, \ \text{ for } v > 0,$$

$$2f'' + g''(s + g' - c) < 0, \ \text{ for } v < 0,$$

$$2f'' + g''(g' - c - s) > 0, \ \text{ for } v > 0,$$

147

$$2f'' + g''(g' - c - s) < 0, \text{ for } v < 0,$$

where $s = \sqrt{(g' - c)^2 + 4f'}$.

Example 11.0.1 *Besides the system of elasticity, there are many other function pairs (f, g) which satisfy the condition (A). For instance, if we choose $f'(v) = (v^2 + d)^l$, $g'(v) - c = k(v^2 + e)^m$, where d, e, m, l are positive constants, k is a non-negative constant and $e \geq d, l \geq m$, then it is easy to check that (A) is satisfied.*

By simple calculations, the two eigenvalues of system (11.0.1) are

$$\lambda_1 = \frac{c + g' - s_2}{2}, \quad \lambda_2 = \frac{c + g' + s_2}{2} \tag{11.0.3}$$

with corresponding right eigenvectors

$$r_1 = (g' - c + s_2, -2)^T, \quad r_2 = (g' - c - s_2, -2)^T. \tag{11.0.4}$$

So

$$\nabla \lambda_1 \cdot r_1 = \frac{2f'' - g''(s_2 - (g' - c))}{s_2},$$
$$\nabla \lambda_2 \cdot r_2 = -\frac{2f'' + g''(s_2 + (g' - c))}{s_2}. \tag{11.0.5}$$

Thus, by the assumption (A), the system (11.0.1) is strictly hyperbolic, but both characteristic fields are linearly degenerate on the line $v = 0$.

Similar to system (10.0.1), the nonlinear flux functions in (11.0.1) are also in implicit forms. We may use the same fashion as given in Chapter 10 to construct the entropy-entropy flux pairs of system (11.0.1) in the following special form:

$$\eta_k^1 = e^{kw}(a_1(v) + \frac{b_1(v, k)}{k}), \quad q_k^1 = e^{kw}(c_1(v) + \frac{d_1(v, k)}{k});$$

$$\eta_{-k}^2 = e^{-kw}(a_2(v) + \frac{b_2(v, k)}{k}), \quad q_{-k}^2 = e^{-kw}(c_2(v) + \frac{d_2(v, k)}{k});$$

$$\eta_{-k}^1 = e^{-kz}(a_3(v) + \frac{b_3(v, k)}{k}), \quad q_{-k}^1 = e^{-kz}(c_3(v) + \frac{d_3(v, k)}{k});$$

$$\eta_k^2 = e^{kz}(a_4(v) + \frac{b_4(v,k)}{k}), \quad q_k^2 = e^{kz}(c_4(v) + \frac{d_4(v,k)}{k}),$$

where w, z are the Riemann invariants of system (11.0.1), i.e.,

$$w = u + \int_0^v \frac{g' - c + s_2}{2} dv, \quad z = u + \int_0^v \frac{g' - c - s_2}{2} dv. \qquad (11.0.6)$$

The necessary estimates for functions $a_i(v), b_i(v, k)$ can be also obtained by the use of the singular perturbation theory of ordinary differential equations of second order (see Lemma 10.2.1). However, a big difference between system (11.0.1) and system (10.0.1) is that the terms $va_i'(v) - a_i(v), i = 1, 2, 3, 4$ could change signs when passing the linearly degenerate line $v = 0$. From the proof in Section 6.4, we can see that some new difficulties will arise from this linear degeneration. The system of quadratic flux given in (6.0.1) is also linearly degenerate, but the common degenerative domain for two characteristic fields is only at the unique point $(u, v) = (0, 0)$. Using the transformation of variable $(su^2 + v^2)$, we can deduce the Young measure, determined by the sequence of viscosity solutions of system (6.0.1), to be a Dirac measure with another support point supposing it is not concentrated in $s = 0$. But to deal with the measure on the line $v = 0$ is more difficult. Some new ideas are introduced in Section 11.3 to reduce the Young to a Dirac measure, and hence to prove the existence of weak solutions to the Cauchy problem (11.0.1), (11.0.2).

We add small, positive perturbation terms to system (11.0.1) and consider the Cauchy problem for the following systems:

$$\begin{cases} u_t + (cu + f(v))_x = \varepsilon u_{xx}, \\ v_t + (u + g(v))_x = \varepsilon v_{xx}, \end{cases} \qquad (11.0.7)$$

with the initial data (11.0.2).

We have the main result in this chapter as follows:

Theorem 11.0.2 *If the initial data $(u_0(x), v_0(x))$ is bounded measurable and the assumption (A) holds, then for any fixed $\varepsilon > 0$, the Cauchy problem (11.0.7), (11.0.2) has a unique global smooth solution $(u^\varepsilon(x, t), v^\varepsilon(x, t))$ satisfying*

$$|u^\varepsilon(x, t)| \le M, \quad |v^\varepsilon(x, t)| \le M, \qquad (11.0.8)$$

where M is a positive constant, independent of ε.

Moreover, there exists a subsequence (still labelled) $(u^\varepsilon(x,t), v^\varepsilon(x,t))$ such that

$$(u^\varepsilon(x,t), v^\varepsilon(x,t)) \to (u(x,t), v(x,t)), \quad a.e. \text{ on } \Omega, \tag{11.0.9}$$

where $\Omega \subset R \times R^+$ is any bounded and open set, $(u(x,t), v(x,t))$ being a weak solution of the Cauchy problem (11.0.1), (11.0.2).

In Section 11.1, we shall prove the existence of viscosity solutions to the Cauchy problem (11.0.7), (11.0.2). In Section 11.2, the method given in Chapter 10 is used to construct the entropy-entropy flux pairs of Lax type, which we shall use in Section 11.2 to prove the existence of weak solutions to the Cauchy problem (11.0.1), (11.0.2).

11.1   Existence of Viscosity Solutions

In this section, we prove the first part in Theorem 11.0.2, namely the existence of global smooth viscosity solutions $(u^\varepsilon, v^\varepsilon)$.

By Theorem 1.0.2, the unique thing is to get the L^∞ estimate (11.0.8).

By the expressions of Riemann invariants of system (11.0.1) given in (11.0.6) and the assumption (A), along the curves in the (u,v)-plane, determined by the equations $w = N, w = -N, z = N$ and $z = -N$, we have the following estimates: On $w = N$, for $v > 0$, there hold

$$\frac{du}{dv} = -\frac{g' - c + s_2}{2} < 0$$

and

$$\frac{d^2u}{dv^2} = -\frac{2f'' + g''(g' - c + s_2)}{2s_2} < 0;$$

On $w = -N$, for $v < 0$, there hold

$$\frac{du}{dv} = -\frac{g' - c + s_2}{2} < 0$$

and

$$\frac{d^2u}{dv^2} = -\frac{2f'' + g''(g' - c + s_2)}{2s_2} > 0;$$

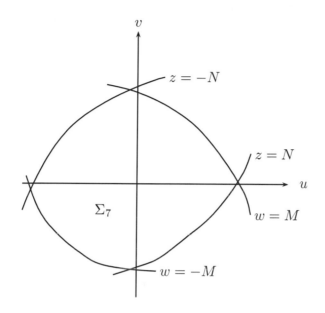

FIGURE 11.1

On $z = N$, for $v > 0$, there hold

$$\frac{du}{dv} = -\frac{g' - c - s_2}{2} > 0$$

and

$$\frac{d^2u}{dv^2} = \frac{2f'' + g''(g' - c - s_2)}{2s_2} < 0;$$

On $z = -N$, for $v > 0$, there hold

$$\frac{du}{dv} = -\frac{g' - c - s_2}{2} > 0,$$

and

$$\frac{d^2u}{dv^2} = \frac{2f'' + g''(g' - c - s_2)}{2s_2} > 0.$$

Therefore

$$\Sigma_7 = \big\{(u, v) : w \leq N, \quad w \geq -N, \quad z \leq N, \quad z \geq -N\big\}$$

is an invariant region (see Figure 11.1), which implies the L^∞ estimate (11.0.8) and hence the existence of the viscosity solutions to the Cauchy problem (11.0.7), (11.0.2).

11.2 Entropy-Entropy Flux Pairs of Lax Type

To prove the second part in Theorem 11.0.2, in this section, we shall construct the entropy-entropy flux pairs of Lax type to system (11.0.1).

From the definition about entropy-entropy flux pair given in Chapter 4, a pair (η, q) of real-valued maps is an entropy-entropy flux pair of system (11.0.1) if all smooth solutions satisfy

$$(c\eta_u + \eta_v, f'\eta_u + g'\eta_v) = (q_u, q_v). \tag{11.2.1}$$

Eliminating the q from (11.2.1), we have

$$f'\eta_{uu} + (g' - c)\eta_{uv} - \eta_{vv} = 0. \tag{11.2.2}$$

Similar to Chapter 10, we can construct four sets of Lax entropy pairs as follows:

$$\eta_k^1 = e^{kw}(a_1(v) + \frac{b_1(v,k)}{k}), \quad q_k^1 = \lambda_2\eta_k^1 + e^{kw}(\frac{a_1'}{k} + \frac{b_1'(v,k)}{k^2}); \tag{11.2.3}$$

$$\eta_{-k}^2 = e^{-kw}(a_2(v) + \frac{b_2(v,k)}{k}), \quad q_{-k}^2 = \lambda_2\eta_{-k}^2 - e^{-kw}(\frac{a_2'}{k} + \frac{b_2'(v,k)}{k^2}); \tag{11.2.4}$$

$$\eta_k^2 = e^{kz}(a_3(v) + \frac{b_3(v,k)}{k}), \quad q_k^2 = \lambda_1\eta_k^2 + e^{kz}(\frac{a_3'}{k} + \frac{b_3'(v,k)}{k^2}); \tag{11.2.5}$$

$$\eta_{-k}^1 = e^{-kz}(a_4(v) + \frac{b_4(v,k)}{k}), \quad q_{-k}^1 = \lambda_1\eta_{-k}^1 - e^{-kz}(\frac{a_4'}{k} + \frac{b_4'(v,k)}{k^2}), \tag{11.2.6}$$

where $a_i(v), b_i(v,k)(i = 1, 2, 3, 4)$ satisfy

$$s_2 a_1' + \frac{g''(g' - c + s_2) + 2f''}{2s_2}a_1 = 0, \quad a_2(v) = a_1(v), \tag{11.2.7}$$

$$s_2 a_3' + \frac{g''(g' - c - s_2) + 2f''}{2s_2} a_3 = 0, \quad a_4(v) = a_3(v), \tag{11.2.8}$$

and

$$a_1'' + s_2 b_1' + \frac{g''(g' - c - s_2) + 2f''}{2s_2} b_1 + \frac{b_1''}{k} = 0; \tag{11.2.9}$$

$$a_2'' - s_2 b_2' - \frac{g''(g' - c + s_2) + 2f''}{2s_2} b_2 + \frac{b_2''}{k} = 0; \tag{11.2.10}$$

$$a_3'' - s_2 b_3' - \frac{g''(g' - c - s_2) + 2f''}{2s_2} b_3 + \frac{b_3''}{k} = 0; \tag{11.2.11}$$

$$a_4'' + s_2 b_4' + \frac{g''(g' - c + s_2) + 2f''}{2s_2} b_4 + \frac{b_4''}{k} = 0. \tag{11.2.12}$$

From (11.2.7) and (11.2.8) we have that

$$a_1 = a_2 = \exp\left(-\int_0^v \frac{g''(g' - c + s_2) + 2f''}{2s_2^2} dv\right) > 0 \text{ for } v \in [-M, M],$$
$$\tag{11.2.13}$$

and

$$a_3 = a_4 = \exp\left(-\int_0^v \frac{g''(g' - c - s_2) + 2f''}{2s_2^2} dv\right) > 0 \text{ for } v \in [-M, M].$$
$$\tag{11.2.14}$$

Using the arguments in Lemma 10.2.1 in Equations (11.2.9)-(11.2.12), we can get the existence of $a_i (i = 1, 2, 3, 4)$ and the uniform bounded estimates of b_i, b_i' with respect to k. Noticing the assumption (A), $a_i' (i = 1, 2, 3, 4)$ all have only one zero point at $v = 0$.

11.3 Existence of Weak Solutions

In this section, we shall use the entropy-entropy flux pairs of Lax type of system (11.0.1) constructed in Section 11.2 to deduce that the Young measures are Dirac ones, and hence to prove the existence of weak solutions given in the second part of Theorem 11.0.2.

It is easy to check that system (11.0.1) has a strictly convex entropy

$$\eta^\star = \frac{1}{2}u^2 + \int_0^v f(s)ds. \tag{11.3.1}$$

Then in a same fashion as Lemma 10.3.1, we can prove the following lemma:

Lemma 11.3.1 *For any entropy-entropy flux pair $(\eta(u,v), q(u,v))$ of system (11.0.1) given by (11.2.3)-(11.3.6),*

$$\eta(u^\varepsilon, v^\varepsilon)_t + q(u^\varepsilon, v^\varepsilon)_x \text{ is compact in } H^{-1}_{loc}(R \times R^+), \tag{11.3.2}$$

with respect to the viscosity solutions $(u^\varepsilon, v^\varepsilon)$ of the Cauchy problem (11.0.7), (11.0.2).

Lemma 11.3.1 guarantees the measure equation to be true, namely

$$< \nu, \eta^1 >< \nu, q^2 > - < \nu, \eta^2 >< \nu, q^1 >< \nu, \eta^1 q^2 - \eta^2 q^1 > \quad (11.3.3)$$

for any entropy-entropy flux pairs $(\eta^i, q^i)(i = 1, 2)$ of system (11.0.1), which satisfy that $\eta^i(u^\varepsilon, v^\varepsilon)_t + q^i(u^\varepsilon, v^\varepsilon)_x$ is compact in $H^{-1}_{loc}(R \times R^+)$.

Since $a_i > 0$ for all $v \in [-M, M]$, then clearly $< \nu, \eta^1_{\pm k} > > 0$ and $< \nu, \eta^2_{\pm k} > > 0$.

Let Q denote the smallest characteristic rectangle:

$$Q = \{(u,v) : w_- \leq w \leq w_+, \quad z_- \leq z \leq z_+\}.$$

We introduce four new probability measures $\mu^+_k, \mu^-_k, \theta^+_k$ and θ^-_k on Q defined by

$$< \mu^+_k, h >=< \nu, h\eta^1_k > / < \nu, \eta^1_k >, \tag{11.3.4}$$

$$< \mu^-_k, h >=< \nu, h\eta^2_{-k} > / < \nu, \eta^2_{-k} >, \tag{11.3.5}$$

$$< \theta^+_k, h >=< \nu, h\eta^2_k > / < \nu, \eta^1_k > \tag{11.3.6}$$

and

$$< \theta^-_k, h >=< \nu, h\eta^1_{-k} > / < \nu, \eta^2_{-k} >, \tag{11.3.7}$$

where $h = h(u, v)$ denotes an arbitrary continuous function. Clearly $\mu_k^+, \mu_k^-, \theta_k^+$ and θ_k^- all are uniformly bounded with respect to k. Then as a consequence of weak-star compactness, there exist probability measures $\mu^\pm$ and $\theta^\pm$ on Q such that

$$< \mu^\pm, h >= \lim_{k \to \infty} < \mu_k^\pm, h > \qquad (11.3.8)$$

and

$$< \theta^\pm, h >= \lim_{k \to \infty} < \theta_k^\pm, h > \qquad (11.3.9)$$

after the selection of an appropriate subsequence.

Moreover, similar to the proof in (6.4.8), we have that the measures $\mu^+, \mu^-, \theta^+, \theta^-$ are respectively concentrated on the boundary sections of Q associated with w and z, i.e.,

$$\text{supp } \mu^+ = Q \bigcap \{(u, v) : w = w_+\} = I_w^+, \qquad (11.3.10)$$

$$\text{supp } \mu^- = Q \bigcap \{(u, v) : w = w_-\} = I_w^-, \qquad (11.3.11)$$

$$\text{supp } \theta^+ = Q \bigcap \{(u, v) : z = z_+\} = I_z^+, \qquad (11.3.12)$$

and

$$\text{supp } \theta^- = Q \bigcap \{(u, v) : z = z_-\} = I_z^-. \qquad (11.3.13)$$

Similar to the proof of (6.4.15), we have

$$< \mu^+, q - \lambda_2 \eta >< \mu^-, q - \lambda_2 \eta > \qquad (11.3.14)$$

and

$$< \theta^+, q - \lambda_1 \eta >< \theta^-, q - \lambda_1 \eta > \qquad (11.3.15)$$

for any (η, q) satisfying that $\eta_t + q_x$ is compact in $H_{loc}^{-1}(R \times R^+)$.

Now we are in the position to prove that the Young measure ν is the Dirac measure.

Let the line $v = 0$ in the (u, v)-plane be Γ, i.e,

$$\Gamma = \{(u, v) : v = 0\} = \{(w, z) : w = z\}.$$

Step 1. If any one of two points $P^+ = I_w^+ \cap I_z^+$ and $P^- = I_w^- \cap I_z^-$ does not lie on Γ, then it is easy to deduce that the Young measure is a Dirac measure.

In fact, in this case, at least one of the boundary arcs $I_w^+, I_w^-, I_z^+, I_z^-$ is contained in the component of Γ^c, i.e., in an open set where both of the characteristic fields are genuinely nonlinear, or all $a_i'(v), i = 1, 2, 3, 4$ are not zero. For example, if P^+ lies in Γ^c, then either $I_w^+ \subset \Gamma^c$ or $I_z^+ \subset \Gamma^c$.

If $I_w^+ \subset \Gamma^c$, we may use the same method given in the proof of (6.4.16) and (6.4.17) to deduce that $w^+ = w^-$ if we go by (η_k^1, q_k^1) instead of (η, q) in (11.3.14).

Then the support of ν is reduced to the line $I = I_w^+ = I_w^-$, where both of the characteristic fields are genuinely nonlinear. Thus we can apply the entropy-entropy flux pair (η_k^2, q_k^2) instead of (η, q) in (11.3.15) to deduce $z^+ = z^-$.

Step 2. If the points $P^+ = I_w^+ \cap I_z^+$ and $P^- = I_w^- \cap I_z^-$ both lie on Γ, we observe that the restriction of $a_i'(v), i = 1, 2$ to I_w^+ vanishes at only one point, namely P^+, while the restriction of $a_i'(v), i = 1, 2$ to I_w^- vanishes at only one point, namely P^-. Thus, the support of the boundary measures μ^+ and μ^- are contained within arcs I_w^+ and I_w^- along which $a_i'(v), i = 1, 2$ maintain one sign. Then, μ^+ and μ^- both are Dirac measures and the supports are contained at P^+ and P^-, respectively.

In fact, we argue that if besides the point P^+, the support of measure μ^+ has another point in I_w^+, then we can again apply (η_k^1, q_k^1) instead of (η, q) in (11.3.14) to deduce that $w^+ = w^-$, and hence $P^+ = P^-$. This is a contradiction.

Similarly we can prove that θ^+ and θ^- both are also Dirac measures and the supports are concentrated at P^+ and P^-, respectively.

Therefore

$$\text{supp } \mu^+ = \text{ supp } \theta^+ = P^+, \quad \text{supp } \mu^- = \text{ supp } \theta^- = P^-. \quad (11.3.16)$$

Noticing (11.3.14) and (11.3.15), we have

$$q(P^+) - \lambda_i(P^+)\eta(P^+) = q(P^-) - \lambda_i(P^-)\eta(P^-), \quad (11.3.17)$$

for $i = 1$ and $i = 2$ and for all pairs (η, q).

Let $P^+ = (u^+, 0), P^- = (u^-, 0)$. Then especially choosing $(\eta, q) = (v, u + g(v))$, we have

$$u^+ + g(0) = u^- + g(0)$$

and hence $u^+ = u^-$. Thus the Young measure ν is a Dirac measure, which implies the existence of weak solutions to the Cauchy problem (11.0.1), (11.0.2). ∎

11.4 Related Results

The large data existence theorem of global L^∞ weak solutions for the Cauchy problem (11.0.1), (11.0.2) was first established by DiPerna [Di3] for the system of elasticity, namely $g(v) = c = 0$ in system (11.0.1). This is also the first application of the compensated compactness method on hyperbolic systems of two equations. In DiPerna's original paper, the idea to use the entropy-entropy flux pairs of Lax type to reduce the Young measure to be a Dirac measure was first introduced to this strictly hyperbolic system of elasticity. The proof in this chapter for more general systems (11.0.1) is from [Lu6].

Chapter 12

L^p Case to Systems of Elasticity

In the last chapter, we studied the extended systems of one-dimensional nonlinear elasticity in Lagrangian coordinates

$$\begin{cases} u_t + f(v)_x = 0 \\ v_t + u_x = 0, \end{cases} \qquad (12.0.1)$$

with bounded measurable initial data

$$(u(x,0), v(x,0)) = (u_0(x), v_0(x)), \qquad (12.0.2)$$

where v is the strain, $f(v)$ the stress, and u the velocity. The basic assumptions on the nonlinear function $f(v)$ are as follows:

(a) $f'(v) \geq c > 0$, $\quad (b)$ $v \cdot f''(v) > 0 \ \forall \ v \neq 0$.

The condition (a) ensures the hyperbolicity of the system, which is essential for all existence results we have had up to now for the system of elasticity; and the condition (b) ensures that the system is linearly degenerate only on the line $v = 0$ as well as the L^∞ estimate of solution, which makes the construction of entropy-entropy flux pairs and the reduction of the Young measure to be a Dirac measure much easier. However, if the condition (b) does not hold, in this case, the simplest situation is if the second derivative has still one zero point, namely $v \cdot f''(v) < 0 \ \forall \ v \neq 0$, then system (12.0.1) no longer has the L^∞ estimate although it is still linearly degenerate only on the same

159

line $v = 0$. In this case, only the $L^p(1 < p < \infty)$ a priori estimate is available (cf. [Da1]).

As shown in Chapter 3, for the scalar equation, we can easily extend the compensated compactness method from L^∞ space to L^p space since any smooth function is an entropy for the scalar equation. However, for systems of two or more equations, there are many difficulties in the case of L^p space, such as how to construct suitable entropy-entropy flux pairs (η, q) such that the compactness of $\eta_t(s^l(x,t)) + q_x(s^l(x,t))$ in $W^{-1,2}_{loc}(R \times R^+)$ holds with respect to a suitable sequence of approximated solutions s^l; how to reduce the corresponding Young measures, having unbounded support sets, to be Dirac measures and so on. It is very important both in mathematics and in physics to prove the existence of L^p solutions for hyperbolic systems. Unfortunately, until now, the existence of L^p solutions was obtained only for the above physical model - the system of elasticity. (See [FS1, FS2] for some mathematical systems.)

It is interesting that for the above system of elasticity, at almost the same time, there are two different proofs obtained by Lin [Lin] and Shearer [Sh] independently.

In this chapter, we shall introduce these proofs as well as an application of Shearer's proof on an extended system of elasticity of three equations (12.3.1).

12.1   Lin's Proof for Artificial Viscosity

In this section, we shall introduce a proof of L^p weak solution existence to the Cauchy problem (12.0.1), (12.0.2), which was obtained by Lin by the compensated compactness method coupled with the artificial viscosity approximation.

The artificial viscosity solutions to the Cauchy problem (12.0.1), (12.0.2) satisfy the following singular parabolic systems:

$$\begin{cases} u_t + f(v)_x = \varepsilon u_{xx} \\ v_t + u_x = \varepsilon v_{xx}, \end{cases} \tag{12.1.1}$$

with the initial data (12.0.2).

Our basic assumptions are as follows:

(A_1) There exist positive constants M, M_1 such that

$$f(v) \in C^4(R), \quad |(\frac{d}{dv})^k f(v)| \le M, \; \forall v \in R, k = 2, 3, 4,$$

and $(f'(v))^{-\frac{1}{2}}$ is concave for $v \ge M_1$, convex for $v \le -M_1$ for a large constant M_1.

(A_2) There is a constant $c > 0$ such that $f'(v) \ge c, \; \forall v \in R$.

(A_3) $v \cdot f''(v) < 0, \; \forall v \in R - \{0\}$.

(A_4) There exist real numbers $\bar{u}, \bar{v}, w^0, z^0$ and $w^0 > z^0$ such that

$$u_0(x) - \bar{u} \in L^2(R), \quad v_0(x) - \bar{v} \in L^2(R),$$

$$w(u_0(x), v_0(x)) \ge w^0, \quad z(u_0(x), v_0(x)) \le z^0, \quad \forall x \in R,$$

where w and z are two Riemann invariants of system (12.0.1),

$$w(u, v) - u + \int_0^v (f'(s))^{\frac{1}{2}} ds, \quad z(u, v) = u - \int_0^v (f'(s))^{\frac{1}{2}} ds.$$

Then we have the following theorem:

Theorem 12.1.1 *(P.X. Lin) If the assumptions $(A_1)-(A_4)$ hold, then for any fixed $\varepsilon > 0$, the global smooth solution $(u^\varepsilon(x, t), v^\varepsilon(x, t))$ of the Cauchy problem (12.1.1), (12.0.2) exists. Furthermore, there exist a subsequence (still labelled) $(u^\varepsilon(x, t), v^\varepsilon(x, t))$ and $u(x, t), v(x, t) \in L^\infty([0, \infty), L^2(R))$ such that*

$$(u^\varepsilon(x, t), v^\varepsilon(x, t)) \to (u(x, t), v(x, t)), \quad a.e. \text{ in } \Omega, \qquad (12.1.2)$$

where $\Omega \in R \times R^+$ is any open and bounded set. Therefore the limit pair of functions $(u(x, t), v(x, t))$ is a weak solution of the Cauchy problem (12.0.1), (12.0.2).

Outline of the proof of Theorem 12.1.1. Using the theory of invariant region, we can prove that

$$w(u^\varepsilon(x, t), v^\varepsilon(x, t)) \ge w^0 > z^0 \ge z(u^\varepsilon(x, t), v^\varepsilon(x, t)). \qquad (12.1.3)$$

Then we may construct four families of entropy-entropy flux pairs of Lax type as follows:

$$\eta_{\pm k}(w, z) = e^{\pm kz}(A_0 + A_1(\pm k)^{-1}) + P_{\pm k};$$

$$q_{\pm k}(w, z) = e^{\pm kz}(B_0 + B_1(\pm k)^{-1}) + Q_{\pm k};$$

$$\bar{\eta}_{\pm k}(w, z) = e^{\pm kw}(a_0 + a_1(\pm k)^{-1}) + p_{\pm k};$$

$$\bar{q}_{\pm k}(w, z) = e^{\pm kw}(b_0 + b_1(\pm k)^{-1}) + q_{\pm k},$$

where $A_0(w, z)$, $A_1(w, z)$, $B_0(w, z)$, $B_1(w, z)$ are functions having compact support sets on $z \in [z^-, z^+]$, z^- and z^+ are constants, $z^+ < w^0$; $a_0(w, z)$, $a_1(w, z)$, $b_0(w, z)$, $b_1(w, z)$ are functions having compact support sets on $w \in [w^-, w^+]$, w^- and w^+ are constants, $w^+ > z^0$. Moreover $P_{\pm k}(w, z)$, $Q_{\pm k}(w, z)$, $p_{\pm k}(w, z)$, $q_{\pm k}(w, z)$ have suitable bounds such that the entropies η and corresponding entropy fluxes q satisfy

$$|\nabla^2 \eta| \leq C, \quad |\nabla \eta| \leq C(1 + |u|^\alpha + |v|^\alpha) \qquad (12.1.4)$$

and

$$|\eta| \leq C(1 + |u|^\alpha + |v|^\alpha), \quad |q| \leq C(1 + |u|^\alpha + |v|^\alpha), \qquad (12.1.5)$$

for $0 < \alpha < 1$.

Noticing that system (12.0.1) has a strictly convex entropy

$$\eta^\star = \frac{1}{2}u^2 + \int_0^v f(v)dv$$

and a corresponding entropy flux

$$q^\star = uf(v),$$

we have that

$$\varepsilon^{\frac{1}{2}}\partial_x u^\varepsilon, \quad \varepsilon^{\frac{1}{2}}\partial_x v^\varepsilon \quad \text{are uniformly bounded in } L^2_{loc}(R \times R^+). \quad (12.1.6)$$

Because of the growth conditions (12.1.4)-(12.1.5) on the entropy-entropy flux pairs (η, q), we can apply Theorem 2.3.2 to prove the compactness of $\eta_t(u^\varepsilon, v^\varepsilon) + q_x(u^\varepsilon, v^\varepsilon)$ in $W^{-1,2}_{loc}(R \times R^+)$ with respect to the artificial viscosity approximation $(u^\varepsilon(x, t), v^\varepsilon(x, t))$. Finally, combining some basic ideas given by DiPerna in the L^∞ space (see Section 11.2) with these entropy-entropy flux pairs, through a complicated analysis, we can prove the compactness (12.1.2) and hence the existence of a weak solution of the Cauchy problem (12.0.1), (12.0.2). This completes the proof of Theorem 12.1.1. ∎

More details about the proof of Theorem 12.1.1 can be found in [Lin].

12.2   Shearer's Proof for Physical Viscosity

As shown in the last section, in the proof of Theorem 12.1.1, one basic technical restriction is that the artificial viscosity solutions satisfy the invariant region (12.1.3), which forces us to use artificial viscosity exactly as in (12.1.1).

However, as a physical model, the following physical viscosity approximation to the system of elasticity (12.0.1) is of more interest:

$$\begin{cases} u_t + f(v)_x = \varepsilon u_{xx} \\ v_t + u_x = 0, \end{cases} \tag{12.2.1}$$

in which the invariant region (12.1.3), generally speaking, does not hold.

At almost the same time, independent of Lin's proof, J.W. Shearer (cf. [Sh]) considered the L^p solution for the same system of elasticity (12.0.1) also with the artificial viscosity approximation (12.1.3), but using different entropy-entropy flux pairs, which were extended late by Serre and Shearer to study the compactness of the physical viscosity approximation (12.2.1).

Roughly speaking, Shearer constructed two classes of entropy and entropy flux pairs to system (12.0.1). One class is the Fourier entropy and the other is the half plane supported entropy. Both classes of entropy-entropy flux pairs (η, q) satisfy the following estimates:

$$\begin{cases} \eta(u, v) = (f'(v))^{-\frac{1}{4}} O(1), & q(u, v) = (f'(v))^{\frac{1}{4}} O(1), \\ \eta_u(u, v) = (f'(v))^{-\frac{1}{4}} O(1), & \eta_v(u, v) = (f'(v))^{\frac{1}{4}} O(1), \\ \eta_{uu}(u, v) = (f'(v))^{-\frac{1}{4}} O(1), & \eta_{uv}(u, v) = (f'(v))^{\frac{1}{4}} O(1), \\ \eta_{vv}(u, v) = (f'(v))^{\frac{3}{4}} O(1), \end{cases} \tag{12.2.2}$$

where $O(1)$ denotes a bounded function.

Shearer's result is given in the following theorem:

Theorem 12.2.1 *(J.W. Shearer) If*

$$f'(v) \geq c > 0, \quad f''(v) \neq 0$$

and

$$f''(v) \in L^1 \cap L^\infty(R), \quad f'''(v) \in L^\infty(R),$$

then there exists a subsequence (still labelled) $(u^\varepsilon(x,t), v^\varepsilon(x,t))$ *of the artificial viscosity solutions* $(u^\varepsilon(x,t), v^\varepsilon(x,t))$ *of the Cauchy problem (12.1.1), (12.0.2) and* $u(x,t), v(x,t) \in L^\infty([0,\infty), L^2(R))$ *such that*

$$(u^\varepsilon(x,t), v^\varepsilon(x,t)) \to (u(x,t), v(x,t)), \quad a.e. \ in \ \Omega, \qquad (12.2.3)$$

where $\Omega \in R \times R^+$ *is any open and bounded set. Therefore the limit pair of functions* $(u(x,t), v(x,t))$ *is a weak solution of the Cauchy problem (12.0.1), (12.0.2).*

Remark 12.2.2 *The existence result in Theorem 12.2.1 is almost the same as that in Theorem 12.1.1, but the two classes of entropy pairs constructed by Shearer are more flexible and could be applied to study many different problems as follows:*

(1) The compactness of physical viscosity solutions of the Cauchy problem (12.2.1) with the initial data (12.0.2);

(2) The existence of global L^P weak solutions for the system of adiabatic gas flow through porous media (12.3.1), in which there are three conservation laws;

(3) The relaxation problem to hyperbolic systems with more than two equations.

The application of Shearer's proof on (2) in Remark 12.2.2 is given in the next section and the details about the application on (3) can be found in Chapter 16.

The following is about the application of Shearer's proof on (1) in Remark 12.2.2.

Consider the physical viscosity approximation, that is, when system (12.0.1) is approximated by its singular perturbation, the Cauchy problem of system (12.2.1) with the initial data (12.0.2). Then we have the following compact result obtained by Serre and Shearer (unpublished, cf. [Sh]):

Theorem 12.2.3 *(Serre and Shearer) If*

$$f'(v) \geq c > 0, \quad v \cdot f''(v) < 0, \ \forall \ v \in R - \{0\}$$

and

$$f''(v) \in L^1 \cap L^\infty(R), \quad f'''(v) \in L^\infty(R),$$

then there exists a subsequence (still labelled) $(u^\varepsilon(x,t), v^\varepsilon(x,t))$ of the Cauchy problem (12.2.1), (12.0.2) and $u, v \in L^\infty([0,\infty), L^2(R))$ such that

$$(u^\varepsilon(x,t), v^\varepsilon(x,t)) \to (u(x,t), v(x,t)), \quad a.e. \ in \ \Omega. \tag{12.2.4}$$

Therefore the limit pair of functions $(u(x,t), v(x,t))$ is a weak solution of the Cauchy problem (12.0.1), (12.0.2).

12.3 System of Adiabatic Gas Flow

In this section, we are concerned with the existence of weak solutions of the Cauchy problem for the nonlinear system of three equations:

$$\begin{cases} v_t - u_x = 0, \\ u_t - \sigma(v,s)_x + \alpha u = 0, \\ s_t = 0, \end{cases} \tag{12.3.1}$$

with L^2 bounded initial data

$$(v(x,0), u(x,0), s(x,0)) = (v_0(x), u_0(x), s_0(x)), \tag{12.3.2}$$

where $\alpha \geq 0$ is a constant. System (12.3.1) can be used to model the adiabatic gas flow through porous media, where v is specific volume, u denotes velocity, s stands for entropy, and σ denotes pressure. Its form in Eulerian coordinates is also a model of isothermal unsteady two-phase flow in pipelines (cf. [LLL]).

In dealing with the Cauchy problem (12.3.1), (12.3.2), one basic difficulty is the *a priori* estimate of the viscosity solutions, independent of ε in a suitable L^p space $(p > 1)$, of the following parabolic system:

$$\begin{cases} v_t - u_x = \varepsilon v_{xx} \\ u_t - \sigma(v,s)_x + \alpha u = \varepsilon u_{xx} \\ s_t = \varepsilon s_{xx}, \end{cases} \tag{12.3.3}$$

with the initial data

$$(v_0^\varepsilon(x,0), u_0^\varepsilon(x,0), s_0^\varepsilon(x,0)) = (v_0^\varepsilon(x), u_0^\varepsilon(x), s_0^\varepsilon(x)). \tag{12.3.4}$$

Since system (12.3.1), in general, cannot be diagonalized by Riemann invariants method, it is not to be expected that viscosity solutions $(v^\varepsilon, u^\varepsilon, s^\varepsilon)$ of the Cauchy problem (12.3.3), (12.3.4) will be bounded in L^∞ space, uniformly in ε, by using the invariant region principle [CCS]. We have to search for solutions of system (12.3.1) in L^P space. In some sense, the *a priori* estimate of the solutions of the Cauchy problem (12.3.3), (12.3.4) in L^2 space is easy to get, if we can find a strictly convex entropy for system (12.3.1). However, a new difficulty arises when considering the compactness of the viscosity solutions in L^P space by trying to use the compensated compactness method. Shearer's work given in Section 12.2 provided us with an ideal framework to deal with system (12.3.1) with three conservation laws.

In this section, we shall study the global generalized solution for the Cauchy problem (12.3.1), (12.3.2) by the compensated compactness method combined with the entropy pairs constructed by Shearer.

We make the assumptions about the nonlinear function $\sigma(v, s)$ and the initial data as follows:

(1) $\sigma(v, s) = \sigma(v)g(s) - cs, g(s) \in C^3$, and $\sigma(v)$ satisfies

    (a) $\sigma(v) \in C^3(R), \sigma(0) = 0, \sigma'(v) \geq d > c^2$, for a constant d;

    (b) $\sigma''(0) = 0$, and $\sigma''(v) \neq 0$ for $v \neq 0$;

    (c) $\sigma''(v), \sigma'''(v) \in L^2 \cap L^\infty$,

(2) $(v_0(x), u_0(x), s_0(x))$ are all bounded in L^2 and tend to zero as $|x| \to \infty$ sufficiently fast such that

$$\lim_{|x|\to\pm\infty} (v_0^\varepsilon(x), u_0^\varepsilon(x), s_0^\varepsilon(x)) = (0, 0, 0), \qquad (12.3.5)$$

$$\lim_{|x|\to\pm\infty} \left(\frac{dv_0^\varepsilon(x)}{dx}, \frac{du_0^\varepsilon(x)}{dx}, \frac{ds_0^\varepsilon(x)}{dx} \right) = (0, 0, 0), \qquad (12.3.6)$$

where $(v_0^\varepsilon(x), u_0^\varepsilon(x), s_0^\varepsilon(x))$ are smooth and obtained by smoothing the initial data $(v_0(x), u_0(x), s_0(x))$ with a mollifier, satisfying

$$\lim_{\varepsilon\to 0}(v_0^\varepsilon(x), u_0^\varepsilon(x), s_0^\varepsilon(x))(v_0(x), u_0(x), s_0(x)), \quad a.e., \qquad (12.3.7)$$

$$\|v_0^\varepsilon(x)\|_{L^2} \leq \|v_0(x)\|_{L^2} \leq M, \quad \|v_0^\varepsilon(x)\|_{H^1(R)} \leq M(\varepsilon), \quad (12.3.8)$$

$$\|u_0^\varepsilon(x)\|_{L^2} \le \|u_0(x)\|_{L^2} \le M, \quad \|u_0^\varepsilon(x)\|_{H^1(R)} \le M(\varepsilon), \quad (12.3.9)$$

$$\|s_0^\varepsilon(x)\|_{L^2} \le \|s_0(x)\|_{L^2} \le M,$$
$$(12.3.10)$$
$$\|s_0^\varepsilon(x)\|_{H^1(R)} \le \|s_0(x)\|_{H^1(R)} \le M$$

and

$$\|\frac{d^i v_0^\varepsilon(x)}{dx^i}\|, \quad \|\frac{d^i u_0^\varepsilon(x)}{dx^i}\|, \quad \|\frac{d^i s_0^\varepsilon(x)}{dx^i}\| \le M(\varepsilon), \quad i = 0, 1, 2,$$
$$(12.3.11)$$

where M is a positive constant independent of ε, and $M(\varepsilon)$ a positive constant, but dependent on ε.

We have the main result in the following theorem:

Theorem 12.3.1 *Let the conditions in (1) and (2) hold. Then for any fixed ε, there is a global solution $(v^\varepsilon, u^\varepsilon, s^\varepsilon)$ of the Cauchy problem (12.3.3), (12.3.4) such that the following estimates hold:*

$$\|v^\varepsilon(\cdot, t)\|_{L^2(R \times R^+)}, \quad \|u^\varepsilon(\cdot, t)\|_{L^2(R \times R^+)}, \quad \|s^\varepsilon(\cdot, t)\|_{H^1(R \times R^+)} \le M.$$
$$(12.3.12)$$

Moreover, there exists a subsequence $(v^\varepsilon, u^\varepsilon, s^\varepsilon)$ (still labelled $(v^\varepsilon, u^\varepsilon, s^\varepsilon)$) such that

$$(v^\varepsilon, u^\varepsilon, s^\varepsilon) \to (v, u, s) \quad as\ \varepsilon \to 0, \quad (12.3.13)$$

the limit pair of functions (v, u, s) is bounded in $L^2(R \times R^+)$ and there is a weak solution of the Cauchy problem (12.3.1), (12.3.2).

Proof. To prove the existence of viscosity solutions $(v^\varepsilon, u^\varepsilon, s^\varepsilon)$, it is enough to get the following L^∞ estimate, although the bound $M(\varepsilon, T)$ could tend to infinity as ε tends to zero or the time T tends to infinity:

$$\|v^\varepsilon\| \le M(\varepsilon, T), \quad \|u^\varepsilon\| \le M(\varepsilon, T), \quad \|s^\varepsilon\| \le M(\varepsilon, T). \quad (12.3.14)$$

We multiply the first equation in (12.3.3) by $\sigma(v)g(s) - cs$, the second equation by u, the third equation by $g'(s) \int_0^v \sigma(v)dv - cv + \gamma s$, then add the result to obtain

$$\left(\frac{u^2}{2} + \int_0^v \sigma(v)dvg(s) - csv + \frac{\gamma s^2}{2}\right)_t$$

$$+(csu - \sigma(v)g(s)u)_x + \alpha u^2$$

$$= \varepsilon\left(\frac{u^2}{2} + \int_0^v \sigma(v)dvg(s) - csv + \frac{\gamma s^2}{2}\right)_{xx} \tag{12.3.15}$$

$$-\varepsilon\left(\sigma'(v)g(s)v_x^2 + 2(\sigma(v)g'(s) - c)s_x v_x + u_x^2\right)$$

$$+ \int_0^v \sigma(v)dvg''(s)s_x^2 + \gamma s_x^2\right),$$

where γ is a large positive constant.

By the condition

$$\|s_0(x)\|_{H^1_{loc}(R)} \le M, \quad \lim_{x \to \pm\infty} s_0(x) = 0,$$

we get the uniform boundedness of $s_0(x)$ in $L^\infty(R)$. So the functions $s_0^\varepsilon(x)$ satisfy

$$|s_0^\varepsilon(x)|_{L^\infty} \le M, \quad |\varepsilon^{\frac{1}{2}}s_{0x}^\varepsilon(x)|_{L^\infty} \le M, \quad |\varepsilon s_{0xx}^\varepsilon(x)|_{L^\infty} \le M$$

and hence the third equation in (12.3.3) yields the estimates

$$\begin{cases} \|s^\varepsilon(x,t)\|_{L^\infty} \le M, \quad |\varepsilon^{\frac{1}{2}}s_x^\varepsilon(x,t)|_{L^\infty} \le M, \\[2mm] |\varepsilon s_{xx}^\varepsilon(x,t)|_{L^\infty} \le M, \quad \|s^\varepsilon(\cdot,t)\|_{H^1_{loc}(R)} \le M. \end{cases} \tag{12.3.16}$$

Thus $g(s), g'(s)$ and $g''(s)$ in (12.3.15) all are bounded. Moreover, since

$|\int_0^v \sigma(v)dv| \le Mv^2$, we have by integrating (12.3.15) in $R \times [0, T]$ that,

$$\int_{-\infty}^{\infty} \left(\frac{u^2}{2} + \int_0^v \sigma(v)dvg(s) - csv + \frac{\gamma s^2}{2} \right) dx$$

$$+\varepsilon \int_0^T \int_{-\infty}^{\infty} c_1 v_x^2 + u_x^2 + c_2 s_x^2 + \alpha u^2 dxdt$$

$$\le \int_{-\infty}^{\infty} \left(\frac{u_0^2}{2} + \int_0^{v_0} \sigma(v)dvg(s_0) - cs_0 v_0 + \frac{\gamma s_0^2}{2} \right) dx \qquad (12.3.17)$$

$$+ \int_0^T \int_{-\infty}^{\infty} Mv^2 dxdt \le M_1 + M_2 \int_0^T \int_{-\infty}^{\infty} v^2 dxdt,$$

for some positive constants c_1, c_2, M_1, M_2 and M.

Since

$$\int_0^v \sigma(v)dvg(s) - csv + \frac{\gamma s^2}{2} \ge c_3(v^2 + s^2) \qquad (12.3.18)$$

for a positive constant c_3, we get by applying the Bellman inequality to (12.3.17) that

$$\int_{-\infty}^{\infty} v^2 dx \le M(T), \quad \int_{-\infty}^{\infty} u^2 dx \le M(T), \quad \int_{-\infty}^{\infty} s^2 dx \le M(T),$$

$$(12.3.19)$$

and

$$\varepsilon \int_0^T \int_{-\infty}^{\infty} v_x^2 + u_x^2 + s_x^2 dxdt \le M(T). \qquad (12.3.20)$$

Differentiating the first equation in (12.3.3) with respect to x, then multiplying the result by $2v_x$, we get

$$(v_x)^2 - 2(v_x u_x)_x + 2u_x v_{xx} + 2\varepsilon v_{xx}^2 = \varepsilon(v_x^2)_{xx}. \qquad (12.3.21)$$

Integrating (12.3.21) in $R \times [0, T]$, we have that

$$\int_{-\infty}^{\infty} v_x^2 dx + \int_0^T \int_{-\infty}^{\infty} \varepsilon v_{xx}^2 dxdt \le M(\varepsilon, T). \qquad (12.3.22)$$

Similarly we can get

$$\int_{-\infty}^{\infty} u_x^2 + s_x^2 dx + \int_0^T \int_{-\infty}^{\infty} \varepsilon(u_{xx}^2 + s_{xx}^2) dx dt \leq M(\varepsilon, T). \quad (12.3.23)$$

Now we have the L^∞ estimates given in (12.3.14), and hence the existence of viscosity solutions to the Cauchy problem (12.3.3), (12.3.4). For instance,

$$v^2 = \int_{-\infty}^{x} (v^2)_x dx \leq \int_{-\infty}^{\infty} v^2 dx + \int_{-\infty}^{\infty} v_x^2 dx \leq M(\varepsilon, T).$$

To prove the existence of weak solutions in Theorem 12.3.1, we let s be fixed as a constant and construct the entropy-entropy flux pairs for the following system:

$$\begin{cases} v_t - u_x = 0, \\ u_t - (g(s)\sigma(v))_x = 0. \end{cases} \quad (12.3.24)$$

We make the transformation

$$x = \sqrt{g(s)}y, \quad t = t, \quad v = v, \quad u = \sqrt{g(s)}w, \quad (12.3.25)$$

and so system (12.3.24) is rewritten as follows:

$$\begin{cases} v_t - w_y = 0, \\ w_t - \sigma(v)_y = 0. \end{cases} \quad (12.3.26)$$

We may apply Shearer's method to construct the same two classes of entropy-entropy flux pairs $(\bar{\eta}(v, w), \bar{q}(v, w))$ for the system (12.3.26), one being the Fourier entropy and the other being the half plane supported entropy. Both classes of entropy-entropy flux pairs satisfy the estimates in (12.2.2). Thus we can get the function pairs

$$(\eta(v, u, s), q(v, u, s)) = (\bar{\eta}(v, w), \bar{q}(v, w)) = (\bar{\eta}(v, \frac{u}{\sqrt{g(s)}}), \bar{q}(v, \frac{u}{\sqrt{g(s)}}))$$
$$(12.3.27)$$

to system (12.3.1) satisfying the estimates

$$\begin{cases} \eta(v, u, s) = (\sigma'(v))^{-\frac{1}{4}} O(1), \quad q(v, u, s) = (\sigma'(v))^{\frac{1}{4}} O(1), \\ \eta_u(v, u, s) = (\sigma'(v))^{-\frac{1}{4}} O(1), \quad \eta_v(v, u, s) = (\sigma'(v))^{\frac{1}{4}} O(1), \end{cases}$$
$$(12.3.28)$$

and

$$\begin{cases} \eta_{uu}(v,u,s) = (\sigma'(v))^{-\frac{1}{4}}O(1), \\ \eta_{vu}(v,u,s) = (\sigma'(v))^{\frac{1}{4}}O(1), \\ \eta_{vv}(v,u,s) = (\sigma'(v))^{\frac{3}{4}}O(1), \end{cases} \tag{12.3.29}$$

where $O(1)$ denotes a bounded function. Since

$$(\bar{q}_v, \bar{q}_w) = (-\sigma'(v)\bar{\eta}_w, -\bar{\eta}_v), \tag{12.3.30}$$

we have (consider here s as a variable)

$$\begin{cases} q_v(v,u,s) = -\sigma'(v)\sqrt{g(s)}\eta_u(v,u,s), \\ q_u(v,u,s) = -\dfrac{1}{\sqrt{g(s)}}\eta_v(v,u,s), \\ \eta_s(v,u,s)\bar{\eta}_w\dfrac{\partial w}{\partial s} = \bar{\eta}_w u\Big(\dfrac{1}{\sqrt{g(s)}}\Big)' = -\dfrac{ug'(s)}{g(s)}\eta_u(v,u,s), \\ q_s(v,u,s) = -\dfrac{ug'(s)}{g(s)}q_u(v,u,s). \end{cases} \tag{12.3.31}$$

Multiplying system (12.3.3) by (η_v, η_u, η_s), and using (12.3.31), we have

$$\begin{aligned} \varepsilon\eta_v &v_{xx} + \varepsilon\eta_u u_{xx} + \varepsilon\eta_s s_{xx} \\ &= \eta_v v_t + \eta_u u_t + \eta_s s_t \\ &\quad -\eta_v u_x - \eta_u (g(s)\sigma(v))_x + c\eta_u s_x + \alpha u\eta_u \\ &= \eta_t + q_u\sqrt{g(s)}u_x + c\eta_u s_x + \alpha u\eta_u \\ &\quad +\eta_u\big(g(s)\sigma'(v)v_x + \sigma(v)g'(s)s_x\big) \\ &= \eta_t + \sqrt{g(s)}(q_u u_x + q_v v_x) \\ &\quad -\sigma(v)g'(s)\eta_u s_x + c\eta_u s_x + \alpha u\eta_u \\ &= \eta_t + (\sqrt{g(s)}q)_x - \sqrt{g(s)}q_s s_x - (\sqrt{g(s)})_x q \\ &\quad -\sigma(v)g'(s)\eta_u s_x + c\eta_u s_x + \alpha u\eta_u. \end{aligned} \tag{12.3.32}$$

The left-hand side of (12.3.32),

$$\begin{aligned} \varepsilon\eta_v &v_{xx} + \varepsilon\eta_u u_{xx} + \varepsilon\eta_s s_{xx} \\ &= \varepsilon(\eta_v v_x)_x + \varepsilon(\eta_u u_x)_x + \varepsilon(\eta_s s_x)_x \\ &\quad -\varepsilon\big(\eta_{vv} v_x^2 + \eta_{uu} u_x^2 + \eta_{ss} s_x^2\big) \\ &\quad -\varepsilon\big(2\eta_{vu} v_x u_x + 2\eta_{vs} v_x s_x + 2\eta_{us} u_x s_x\big). \end{aligned} \tag{12.3.33}$$

It follows from (12.3.16), (12.3.19) and (12.3.20) that

$$|\sigma(v)g'(s)\eta_u s_x|_{L^1_{loc}(R \times R^+)} \le M|v|_{L^2_{loc}(R \times R^+)}|s_x|_{L^2_{loc}(R \times R^+)} \le M(T),$$
(12.3.34)

$$\varepsilon|\eta_{ss}s_x^2|_{L^1_{loc}(R \times R^+)} \le \varepsilon M|u^2 s_x^2|_{L^1_{loc}(R \times R^+)} \le M_1|u^2|_{L^1_{loc}(R \times R^+)} \le M(T)$$
(12.3.35)

and

$$\varepsilon|\eta_{vs}v_x s_x + \eta_{us}u_x s_x|_{L^1_{loc}(R \times R^+)}$$

$$\le \varepsilon M(|uv_x s_x| + |us_x u_x|)_{L^1_{loc}(R \times R^+)}$$

$$\le M_1(|u^2|_{L^1_{loc}(R \times R^+)} + \varepsilon|v_x^2|_{L^1_{loc}(R \times R^+)} + \varepsilon|u_x^2|_{L^1_{loc}(R \times R^+)})$$

$$\le M(T),$$
(12.3.36)

for some positive constants $M, M_1, M(T)$.

Therefore we have

$$\eta_t(v, u, s) + (\sqrt{g(s)}q(v, u, s))_x = I_1 + I_2,$$
(12.3.37)

where

$$I_1 = \varepsilon(\eta_v v_x)_x + \varepsilon(\eta_u u_x)_x + \varepsilon(\eta_s s_x)_x \text{ is compact in } H^{-1}_{loc}(R \times R^+),$$
(12.3.38)

and

$$I_2 = (\sqrt{g(s)})_x q + \sqrt{g(s)}q_s s_x + \sigma(v)g'(s)\eta_u s_x$$
$$-c\eta_u s_x - \alpha u\eta_u - \varepsilon(\eta_{vv}v_x^2 + \eta_{uu}u_x^2 + \eta_{ss}s_x^2)$$
$$-\varepsilon(2\eta_{vu}v_x u_x + 2\eta_{vs}v_x s_x + 2\eta_{us}u_x s_x)$$
(12.3.39)

is bounded in $L^1_{loc}(R \times R^+)$ and hence compact in $W^{-1,k}_{loc}(R \times R^+)$ for a constant $k \in (1, 2)$. However, $\eta_t(v, u, s) + (\sqrt{g(s)}q(v, u, s))_x$ is bounded in $W^{-1,p}_{loc}(R \times R^+)$ for $p > 2$ by the estimates in (12.3.28), thus $\eta_t(v, u, s) + (\sqrt{g(s)}q(v, u, s))_x$ is compact in $H^{-1}_{loc}(R \times R^+)$ by Theorem 2.3.2.

From (12.3.16), we have the pointwise convergence of the sequence $\{s^\varepsilon(x,t)\}$. Then the compact support of Young measures $\nu_{(x,t)}$, determined by the viscosity solutions $(v^\varepsilon(x,t), u^\varepsilon(x,t), s^\varepsilon(x,t))$ to the Cauchy problem (12.3.3), (12.3.4) is reduced to the (v,u)-plane, and the following measure equations are satisfied for any entropy-entropy flux pairs $(\eta_i(v,u,s), q_i(v,u,s))$ constructed above:

$$
\begin{aligned}
< \nu, \eta_1(v^\varepsilon, u^\varepsilon, s)\sqrt{g(s)}q_2(v^\varepsilon, u^\varepsilon, s) - \eta_2(v^\varepsilon, u^\varepsilon, s)\sqrt{g(s)}q_1(v^\varepsilon, u^\varepsilon, s) > \\
=< \nu, \eta_1(v^\varepsilon, u^\varepsilon, s) >< \nu, \sqrt{g(s)}q_2(v^\varepsilon, u^\varepsilon, s) > \\
- < \nu, \eta_2(v^\varepsilon, u^\varepsilon, s) >< \nu, \sqrt{g(s)}q_1(v^\varepsilon, u^\varepsilon, s) >,
\end{aligned}
$$

$$(12.3.40)$$

or equivalently

$$
\begin{aligned}
< \nu, \eta_1(v^\varepsilon, u^\varepsilon, s)q_2(v^\varepsilon, u^\varepsilon, s) - \eta_2(v^\varepsilon, u^\varepsilon, s)q_1(v^\varepsilon, u^\varepsilon, s) > \\
=< \nu, \eta_1(v^\varepsilon, u^\varepsilon, s) >< \nu, q_2(v^\varepsilon, u^\varepsilon, s) > \\
- < \nu, \eta_2(v^\varepsilon, u^\varepsilon, s) >< \nu, q_1(v^\varepsilon, u^\varepsilon, s) >
\end{aligned}
$$

$$(12.3.41)$$

since $g(s) > d > 0$.

Now consider s to be a constant in (12.3.41). Then the compactness framework in Theorem 12.2.1 or Theorem 12.2.3 deduces that all the Young measures ν are Dirac measures, which implies the proof of Theorem 12.3.1. ∎

12.4 Related Results

Besides the physical model (12.0.1), the existence of L^p weak solutions is also obtained by Frid and Santos [FS1, FS2] for the following mathematical system of two equations:

$$z_t - (\bar{z}^\gamma)_x = 0, \quad 1 < \gamma < 2, \tag{12.4.1}$$

where $z = u + iv \in \mathcal{C}$.

System (12.3.1) is more or less similar to system (12.0.1), but it is the unique system of three equations for which we can obtain L^p weak solutions until now. System (12.3.1) and the system of chromatography (7.5.1) are only two applications of the compensated compactness method on hyperbolic systems of more than two equations.

The proof of Theorem 12.3.1 is from the paper [LK1].

Chapter 13

Preliminaries in Relaxation Singularity

We are concerned with the system of partial differential equation in the form

$$U(x,t)_t + F(U)_x + \frac{1}{\tau}R(U) = \varepsilon U_{xx}. \qquad (13.0.1)$$

Here $U = U(x,t)$, which takes on value in R^N, represents the density vector of basic physical variables over the space variable x, the quantity τ is the relaxation time, which is small in many physical situations. In the kinetic theory it is the mean free path, in elasticity the duration of memory. ε is the artificial viscosity parameter, or the diffusion coefficient.

When $\varepsilon = 0$, and the corresponding systems

$$U(x,t)_t + F(U)_x = 0 \qquad (13.0.2)$$

are hyperbolic, that is, the $N \times N$ matrix $\nabla F(U)$ has N real eigenvalues, the relaxation systems

$$U(x,t)_t + F(U)_x + \frac{1}{\tau}R(U) = 0 \qquad (13.0.3)$$

in the level of hyperbolic equations arise in many physical situations, such as combustion theory [Lu3, Lu7, Ma], multiphase and phase transition [ChL], chromatography [RAA1, Wh], viscoelasticity [CLL, Na],

175

kinetic theory [BR, Ca, Ce], river flows [KL2, Wh] and traffic flows [Sc, Wh].

Given a system in the form of (13.0.1), an interesting thing both in physics and mathematics is the limit behavior of the solutions $U^{\tau,\varepsilon}(x,t)$ of system (13.0.1) as the relaxation time and the viscosity ε go to zero.

As an illustrative 2×2 model, consider the following system of two equations:

$$\begin{cases} u_t - v_x = \varepsilon u_{xx}, \\ v_t - cu_x + \dfrac{v - h(u)}{\tau} = \varepsilon v_{xx}, \end{cases} \qquad (13.0.4)$$

where c is a constant and $h(u)$ is the equilibrium value for v. System (13.0.4) could be the simplest model of systems in the form (13.0.1).

For system (13.0.4), we shall see that if $\varepsilon = 0$ and the system

$$\begin{cases} u_t - v_x = 0, \\ v_t - cu_x = 0, \end{cases} \qquad (13.0.5)$$

is hyperbolic, i.e., $c > 0$, the basic condition

$$-\sqrt{c} < h'(u) < \sqrt{c}, \qquad (13.0.6)$$

so called the subcharacteristic condition (cf. [Liu]), is necessary to ensure the stability of solutions (u^τ, v^τ) of the following system:

$$\begin{cases} u_t - v_x = 0, \\ v_t - cu_x + \dfrac{v - h(u)}{\tau} = 0 \end{cases} \qquad (13.0.7)$$

as $\tau \to 0$. However if $\varepsilon = 0$ and $c < 0$, system (13.0.4) or system (13.0.7) is ill posed; if $\varepsilon > 0$ and $\tau = o(\varepsilon)$, that is, τ is smaller than ε, then the solutions $(u^{\tau,\varepsilon}, v^{\tau,\varepsilon})$ of system (13.0.4) are always stable.

These can be seen through an asymptotic expansion of the Chapman-Enskog type.

Let

$$v = h(u) + \tau v_1 + O(\tau^2). \qquad (13.0.8)$$

Then from the second equation in (13.0.4) we obtain

$$\begin{aligned} v_1 &= \varepsilon v_{xx} - v_t + cu_x + O(\tau) \\ &= \varepsilon v_{xx} - h'(u)u_t + cu_x + O(\tau) \\ &= \varepsilon v_{xx} - \varepsilon h'(u)u_{xx} - h'(u)v_x + cu_x + O(\tau) \\ &= \varepsilon v_{xx} - \varepsilon h'(u)u_{xx} - (h'(u))^2 u_x + cu_x + O(\tau). \end{aligned} \qquad (13.0.9)$$

Substituting (13.0.9) into the first equation in (13.0.4), we have

$$
\begin{aligned}
u_t - h(u)_x \;&=\; \epsilon u_{xx} + \tau v_{1x} + O(\tau^2) \\
&=\; \Big((\varepsilon + \tau(c - h'(u)^2)) u_x \Big)_x + O(\tau^2) + O(\tau\epsilon).
\end{aligned}
$$
$$(13.0.10)$$

Thus if $\varepsilon = 0, c > h'^2(u)$ or $\varepsilon > 0$ and $\tau = o(\varepsilon)$, in the latter case, $\varepsilon > \tau(h'^2(u) - c)$, Equation (13.0.10) is well posed since the coefficient of the diffusion term is positive. Unfortunately if $\varepsilon = 0$ and $c < 0$, we have the following system:

$$
\begin{cases}
u_t - v_x = 0, \\
v_t + u_x = 0,
\end{cases}
$$
$$(13.0.11)$$

and hence system (13.0.7) is ill posed as we proceed to show.

By the first equation in (13.0.7) there must exist a function w such that

$$
w_x = u, \qquad w_t = v.
$$
$$(13.0.12)$$

Thus the second equation in (13.0.7) can be put in the form

$$
w_{tt} - c w_{xx} + \frac{1}{\tau}(w_t - h(w_x)) = 0.
$$
$$(13.0.13)$$

This is an elliptic equation and its Dirichlet problem in the domain $t > 0$ can be solved with the data

$$
w(x, 0) = \int_0^x u_0(s)\,ds.
$$
$$(13.0.14)$$

But then $v_0(x)$ cannot be chosen independently of $u_0(x)$ since in this case we must have $v_0(x) = w_t(0, x)$ and w depends on $u_0(x)$.

The above analysis illustrates that, besides the subcharacteristic condition (13.0.6) for the stability of solutions of hyperbolic system (13.0.7), viscosity ε in (13.0.1) is not only of mathematical expedience when acting together with relaxation but may also be another necessary stability mechanism.

In Chapter 14, we are concerned with singular limits of stiff relaxation and dominant diffusion for general 2×2 nonlinear systems of conservation laws, that is, the relaxation time τ tends to zero faster than the diffusion parameter ϵ, $\tau = o(\epsilon), \varepsilon \to 0$. Some compactness frameworks without the subcharacteristic condition are established.

In Chapter 15, we shall study the limiting behavior of some special hyperbolic systems of two equations with stiff relaxation terms, where all compact results are based on the subcharacteristic condition.

In Chapter 16, relaxation problems for some special hyperbolic systems of three equations are studied.

Chapter 14

Stiff Relaxation and Dominant Diffusion

In this chapter, we are concerned with singular limits of stiff relaxation and dominant diffusion for general 2×2 nonlinear systems of conservation laws, that is, the relaxation time τ tends to zero faster than the diffusion parameter ϵ, $\tau = o(\epsilon), \epsilon \to 0$. We establish the following general framework: If there exists an *a priori* uniform L^∞ bound with respect to ϵ for the solutions of a system, then the solution sequence converges to the corresponding equilibrium solution of this system. Our results indicate that the convergent behavior of such a limit is independent of either the stability criterion or the hyperbolicity of the corresponding inviscid quasilinear systems, which is not the case for other types of limits in Chapter 15 for relaxation limits without viscosity. This framework applies to some important nonlinear systems with relaxation terms, such as the system of elasticity, the system of isentropic fluid dynamics in Eulerian coordinates, and the extended models of traffic flows. The singular limits are also considered for some physical models, without L^∞ bounded estimates, including the system of isentropic fluid dynamics in Lagrangian coordinates and the models of traffic flows with stiff relaxation terms. The convergence of solutions in L^p to the equilibrium solutions of these systems is established, provided that the relaxation time τ tends to zero faster than ε.

14.1   Compactness Results

We are concerned with singular limits of stiff relaxation and dominant diffusion for the Cauchy problem of general 2×2 quasilinear conservation laws with relaxation and diffusion:

$$\begin{cases} v_t + f(v, u)_x = \varepsilon v_{xx}, \\[2mm] u_t + g(v, u)_x + \frac{1}{\tau} \alpha(v, u)(u - h(v)) = \varepsilon u_{xx}, \end{cases} \qquad (14.1.1)$$

with initial data

$$(v, u)|_{t=0} = (v_0(x), u_0(x)). \qquad (14.1.2)$$

The second equation in (14.1.1) contains a relaxation mechanism, with $h(v)$ as the equilibrium value for u, τ the relaxation time, $\alpha(v, u) > 0$, and ε is the diffusion coefficient. The relaxation term serves as a damping in some suitable system coordinates.

The singular limit problem for (14.1.1) can be considered as a singular perturbation problem as τ tends to zero.

When $\tau = \varepsilon$, the relaxation systems have been studied for some typical models (cf. [ES, Fi, RSK] and the references cited therein).

The relaxation systems will be studied in Chapter 15 in the case where $\varepsilon = o(\tau)$ and the corresponding 2×2 systems

$$\begin{cases} v_t + f(v, u)_x = 0, \\[2mm] u_t + g(v, u)_x = 0, \end{cases} \qquad (14.1.3)$$

are hyperbolic.

In this section, we consider the case of stiff relaxation and dominant diffusion, that is, $\tau = o(\epsilon)$, as $\varepsilon \to 0$.

When the solutions of the Cauchy problem (14.1.1), (14.1.2) are uniformly bounded in L^∞, we show that the limit is always stable and no oscillation arises for any C^1 flux functions f and g.

Theorem 14.1.1 *Let $f, g \in C^1(R^2)$, $h \in C^2(R)$ and $\alpha_0 \leq \alpha \in C(R^2)$ for a positive constant α_0. Let $\tau = o(\varepsilon)$ as $\varepsilon \to 0$. If the solutions $(v^\varepsilon, u^\varepsilon) \equiv (v^{\varepsilon, \tau(\varepsilon)}, u^{\varepsilon, \tau(\varepsilon)})$ of the Cauchy problem (14.1.1), (14.1.2) have an a priori L^∞ bound:*

$$|(v^\varepsilon, u^\varepsilon)(x, t)| \leq M(T), \quad (x, t) \in R \times [0, T], \qquad (14.1.4)$$

for any given time T, where M(T) is independent of ε, then there exists a subsequence $(v^{\varepsilon_k}, u^{\varepsilon_k})$ *converging strongly to the functions* (v, u) *as* $\varepsilon_k \to 0$, *which are the equilibrium states uniquely determined by* (E_1)-(E_2):

(E_1) $u(x, t) = h(v(x, t))$, *for almost all* $(x, t) \in R \times (0, T]$;

(E_2) $v(x, t)$ *is the* L^∞ *entropy solution of the Cauchy problem*

$$v_t + f(v, h(v))_x = 0, \quad v|_{t=0} = v_0(x). \tag{14.1.5}$$

If the solutions of the Cauchy problem (14.1.1)-(14.1.2) have no *a priori* L^∞ estimate, we can prove that the limit is also stable, provided that the functions f, g, and α satisfy certain growth conditions.

Assume

$$(v_0(x) - \bar{v}, u_0(x) - \bar{u}) \in L^\infty \cap L^p(R), \quad \bar{u} = h(\bar{v}), \quad 1 < p < \infty. \tag{14.1.6}$$

Also assume the following conditions:

(B_1) $|f(v, u)| + |g(v, u)| \le c_1 + c_2(|v|^q + |u|^q), \quad q \in [1, 3),$

$\quad |\nabla_{v,u} f(v, u)| + |\nabla_{v,u} g(v, u)| \le c_3 + c_4(|v|^{q-1} + |u|^{q-1});$

(B_2) $0 < c_5 + c_6(|v|^r + |u|^r) \le \alpha(v, u) \le c_7 + c_8(|v|^r + |u|^r), \quad 0 \le r < 4;$

(B_3) $|h(v)| \le c_9 + c_{10}|v|^k, \quad k \ge 1,$

where $q, r, k, c_i, 1 \le i \le 10$, are positive constants.

Theorem 14.1.2 *(I) Assume conditions* (B_1)-(B_3) *are satisfied, and there exists a strictly convex function* $\bar{p}(v, u)$ *in the range of solutions of the Cauchy problem (14.1.1), (14.1.2) with initial data* (v_0, u_0) *satisfying (14.1.6) such that*

$$\bar{p}_u(v, u) = p_1(v, u)(u - h(v)), \quad \bar{p}(v, u) = \eta(v, u) + p(v, u),$$

where $p_1(v, u) \ge c_0 > 0$, $p(v, u)$ *is a smooth function whose second order derivatives are bounded, and* $\eta(v, u)$ *is an entropy of system (14.1.3). Then, if* $2k(q - 1) \le r$ *and* $M_1 \tau \le \varepsilon$ *for some large constant* M_1, *for fixed* ε, τ, *the Cauchy problem (14.1.1), (14.1.2) with initial data*

(v_0, u_0) satisfying (14.1.6) has a unique smooth solution $(v^{\varepsilon,\tau}, u^{\varepsilon,\tau})$ satisfying

$$|(v^{\varepsilon,\tau}, u^{\varepsilon,\tau})(x,t)| \le C(\varepsilon, \tau), \qquad (14.1.7)$$

for some constant $C(\varepsilon, \tau) > 0$ depending on ε and τ.

(II) If the conditions of (I) are satisfied and $2k(q-1) + p - 2 \le r$, then $|v^{\varepsilon,\tau} - \bar{v}|_p \le M$.

(III) If the conditions of (II) are satisfied, $p > max\{1, q, qk\}$, and $\tau = o(\varepsilon)$ as $\varepsilon \to 0$, then there exists a subsequence $(v^{\varepsilon k}, u^{\varepsilon k})$ of $(v^\varepsilon, u^\varepsilon) \equiv (v^{\varepsilon,\tau(\varepsilon)}, u^{\varepsilon,\tau(\varepsilon)})$, converging pointwisely almost everywhere:

$$(v^{\varepsilon k}, u^{\varepsilon k}) \to (v, u), \quad as \quad \varepsilon_k \to 0, \qquad (14.1.8)$$

where the limit functions (v, u) satisfy (F_1)-(F_2):

(F_1) $u(x,t) = h(v(x,t))$, for almost all $(x,t) \in R \times (0,T]$;

(F_2) $v(x,t)$ is the unique L^p entropy solution of the Cauchy problem:

$$v_t + f(v, h(v))_x = 0, \quad v|_{t=0} = v_0(x). \qquad (14.1.9)$$

Remark 14.1.3 If $2k(q-1) \le r$, then from $(B_1) - (B_3)$, we have

(B_4) $\{f_u(v, \beta)^2 + g_u(v, \beta)^2 + f'(v, h(v))^2 + g'(v, h(v))^2\}/\alpha(v, u) \le M$,

where β takes a value between u and $h(v)$.

This inequality will be used to prove Theorem 14.1.2 in Section 14.4.

The proof of Theorem 14.1 is given in Section 14.2. Its applications to the system of elasticity, the system of isentropic gas dynamics in Eulerian coordinates, and the extended models of traffic flows with stiff relaxation terms will be given in Section 14.3.

The proof of Theorem 14.1.2 is given in Section 13.4, whose applications to the system of isentropic gas dynamics in Lagrangian coordinates and the models of traffic flows with stiff relaxation terms are given in Section 14.5.

14.2 Proof of Theorem 14.1.1

Throughout this Chapter, we denote $R_+^2 = R \times R^+ = (-\infty, \infty) \times (0, \infty)$, and $M(T)$ and M are generic constants independent of ϵ and τ, which may be different at each occurrence. Before proving Theorem 14.1.1, we first introduce two lemmas.

First, we have the following global existence result about the Cauchy problem (14.1.1), (14.1.2):

Lemma 14.2.1 *If the solutions of (14.1.1), (14.1.2) have an a priori L^∞ bound (14.1.7), then, for any fixed ε and τ, the Cauchy problem (14.1.1), (14.1.2) has a unique smooth solution $(v^{\varepsilon,\tau}, u^{\varepsilon,\tau})$ on $R \times (0, T]$.*

The local existence and regularity of solutions for $t > 0$ of the Cauchy problem (14.1.1), (14.1.2) can be obtained by applying the Banach contraction mapping theorem to an integral representation of (14.1.1), where the local time depends only on ϵ, τ, the L^∞ norm of the initial data. The global existence is based on the local existence and the *a priori* L^∞ estimate (14.1.7).

Second, we have the following estimates:

Lemma 14.2.2 *If the solutions of (14.1.1)-(14.1.2) have an a priori L^∞ bound (14.1.7) and $f, g \in C^1(R^2), h \in C^2(R)$, then*

$$\|(\varepsilon v_x^2, \varepsilon u_x^2, \frac{(u - h(v))^2}{\tau})\|_{L_{loc}^1(R_+^2)} \leq M, \tag{14.2.1}$$

provided that $M_1\tau \leq \varepsilon$ for some large constant $M_1 > 0$.

Proof. Since (v, u) is bounded, we can choose a large constant C_1 such that the function $p(v, u) = \frac{u^2}{2} - h(v)u + \frac{C_1 v^2}{2}$ satisfies

$$p_{vv}(v, u)v_x^2 + 2p_{vu}(v, u)v_x u_x + p_{uu}(v, u)u_x^2 \geq C_2(v_x^2 + u_x^2), \tag{14.2.2}$$

for some constant $C_2 > 0$.

Multiplying system (14.1.1) by (p_v, p_u), we have from (14.2.2) that

$$p(v, u)_t + p_v(v, u)f(v, u)_x + p_u(v, u)g(v, u)_x$$
$$+ \alpha(v, u)\frac{(u - h(v))^2}{\tau} + \varepsilon C_2(v_x^2 + u_x^2) \tag{14.2.3}$$
$$\leq \varepsilon p_{xx}(v, u).$$

Notice that

$$
\begin{aligned}
p_v(v,u)f(v,u)_x &= (p_v(v,u)(f(v,u) - f(v,h(v))))_x \\
&\quad + p_v(v,h(v))f(v,h(v))_x \\
&\quad - p_{vx}(v,u)(f(v,u) - f(v,h(v))) \\
&\quad + (p_v(v,u) - p_v(v,h(v)))f(v,h(v))_x \\
&= (p_v(v,u)(f(v,u) - f(v,h(v))))_x \\
&\quad + \left(\int^v p_v(s,h(s))f'(s,h(s))ds \right)_x \\
&\quad - (p_{vu}(v,u)u_x + p_{vv}(v,u)v_x)f_u(v,\beta_1)(u - h(v)) \\
&\quad + p_{vu}(v,\beta_2)(u - h(v))f'(v,h(v))v_x
\end{aligned}
\tag{14.2.4}
$$

and

$$
\begin{aligned}
p_u(v,u)g(v,u)_x &= (p_u(v,u)(g(v,u) - g(v,h(v))))_x \\
&\quad + \left(\int^v p_u(s,h(s))g'(s,h(s))ds \right)_x \\
&\quad - (p_{uu}(v,u)u_x + p_{vu}(v,u)v_x)g_u(v,\beta_3)(u - h(v)) \\
&\quad + p_{uu}(v,\beta_4)(u - h(v))g'(v,h(v))v_x,
\end{aligned}
\tag{14.2.5}
$$

where

$$
p_v(v,h(v)) = p_v(v,u)|_{u=h(v)}, \quad f'(v,h(v)) = \frac{df(v,h(v))}{dv}
$$

and $\beta_i, 1 \le i \le 4$, take values between u and $h(v)$.

It follows from (14.2.2)–(14.2.5) that

$$
\begin{aligned}
p(v,u)_t + q(v,u)_x &+ c_0 \frac{(u - h(v))^2}{2\tau} + \varepsilon C_2(v_x^2 + u_x^2) \\
&- \tau C_3(v_x^2 + u_x^2) \\
&\le \varepsilon p_{xx}(v,u),
\end{aligned}
\tag{14.2.6}
$$

for a function q and positive constants c_0 and C_3 depending on the bounds of second derivatives of p and first derivatives of f and g.

Multiplying (14.2.6) by a suitable nonnegative test function and then integrating by parts on R_+^2, we get the estimates in (14.2.1) provided that $2\tau C_3 \le \varepsilon C_2$.

Proof of Theorem 14.1.1. To prove Theorem 14.1.1, we rewrite the first equation in (14.1.1) as follows:

$$v_t + f(v, h(v))_x = \varepsilon v_{xx} + \big(f(v, h(v)) - f(v, u)\big)_x. \qquad (14.2.7)$$

Let $\eta(v)$ be any entropy of the scalar equation

$$v_t + f(v, h(v))_x = 0.$$

Multiplying (14.2.7) by $\eta'(v)$, we have

$$
\begin{aligned}
\eta(v)_t + q(v)_x &= -\eta'(v)(f(v, u) - f(v, h(v)))_x + \varepsilon \eta'(v) v_{xx} \\
&= -(\eta'(v)(f(v, u) - f(v, h(v))))_x + \varepsilon \eta(v)_{xx} \\
&\quad + (f(v, u) - f(v, h(v)))\eta''(v) v_x - \varepsilon \eta''(v) v_x^2 \quad (14.2.8) \\
&= -(\eta'(v) f_u(v, \beta_1)(u - h(v)))_x + \varepsilon \eta(v)_{xx} \\
&\quad + f_u(v, \beta_2)\eta''(v)(u - h(v)) v_x - \varepsilon \eta''(v) v_x^2,
\end{aligned}
$$

where $\beta_i, i = 1, 2$, take values between u and $h(v)$.

It follows from the estimates in (14.2.1) that, on any compact $\Omega \subset R_+^2$, there hold

$$
\int_\Omega |f_u \eta''(v)(u - h(v)) v_x| dx dt
$$
$$
\leq M \left(\int_\Omega \frac{(u - h(v))^2}{\tau} dx dt \right)^{\frac{1}{2}} \left(\int_\Omega \tau v_x^2 dx dt \right)^{\frac{1}{2}} \to 0, \qquad (14.2.9)
$$

and

$$
\left| \int_\Omega (\eta'(v) f_u(u - h(v)))_x \phi \, dx dt \right|
$$
$$
= \left| \int_\Omega \eta'(v) f_u(u - h(v)) \phi_x \, dx dt \right| \qquad (14.2.10)
$$
$$
\leq M \left(\int_\Omega \tau \phi_x^2 dx dt \right)^{\frac{1}{2}} \left(\int_\Omega \frac{(u - h(v))^2}{\tau} dx dt \right)^{\frac{1}{2}} \to 0
$$

as $\varepsilon \to 0$. Moreover, since $\varepsilon \eta''(v) v_x^2$ is bounded in L_{loc}^1 and $\varepsilon \eta(v)_{xx} \to 0$ in the sense of distributions, the right-hand side of (14.2.7) is compact in $W_{loc}^{-1,q}$, for a constant $q \in (1, 2)$. Noticing the left-hand side of (14.2.7) is bounded in $W^{-1,\infty}$, we have that

$$
\eta(v^\varepsilon)_t + q(v^\varepsilon)_x \quad \text{is compact in } W_{loc}^{-1,2}
$$

for any entropy η, with respect to the viscosity solutions v^ε. Therefore the strong convergence of v^ε follows from the compactness framework about the scalar equation given in Chapter 3, which implies the strong convergence of u^ε by the third estimate in (14.2.1). This completes the proof of Theorem 14.1.1. ∎

Remark 14.2.3 *In Theorem 14.1.1, the condition (3.1.6) on the non-linear flux function $f(v, h(v))$ is not assumed to ensure the strong convergence of v^ε. In fact, this condition is removed by Szepessy [Sz] as shown in the last part of Section 3.3 (see [Lu9] for the details).*

14.3   Applications of Theorem 14.1.1

In this section we apply Theorem 14.1.1 to some important physical models such as the system of elasticity, the isentropic system of gas dynamics in Eulerian coordinates, and the extended models of traffic flows with relaxation terms.

14.3.1. The System of Elasticity

The system of elasticity is given by

$$\begin{cases} v_t + u_x = 0, \\ u_t + \sigma(v)_x = 0, \end{cases} \qquad (14.3.1)$$

which describes the balance of mass and momentum. The existence of L^∞ weak solutions for system (14.3.1) with large bounded initial data is obtained in Chapter 11 for the case of $\sigma''(v)v > 0$, for all $v \in R \backslash \{0\}$, and the L^p weak solutions are introduced in Chapter 12 for the case of $\sigma''(v)v < 0$, for all $v \in R \backslash \{0\}$.

Consider the viscosity solutions for the system of elasticity with a relaxation term:

$$\begin{cases} v_t + u_x = \varepsilon v_{xx}, \\ u_t + \sigma(v)_x + \frac{u - h(v)}{\tau} = \varepsilon u_{xx}, \end{cases} \qquad (14.3.2)$$

with bounded initial data $(v_0(x), u_0(x))$.

The zero relaxation and dissipation limits for system (14.3.1) was first studied by Chen, Levermore and Liu ([CLL]). By applying the invariant region arguments, they first obtained the L^∞ estimate of solutions $(v^{\varepsilon,\tau}, u^{\varepsilon,\tau})$ for the Cauchy problem (14.3.2) with the initial data $(v_0(x), u_0(x))$ satisfying following assumptions:

(C_1) $\sigma \in C^2$, $\sigma'(v) > 0$ for all $v \in R$; $v\sigma''(v) > 0$ for all $v \in R\backslash\{0\}$;

(C_2) σ and h satisfy the stability condition $|h'(v)| \le \sqrt{\sigma'(v)}$;

(C_3) $h \in C^1$, $h(v) = \bar{h}$ ($\bar{h}$ a constant), as $|v| \ge M_0$ for a suitable large constant M_0.

In fact, assumption (C_3) can be removed by applying the comparison principle (cf. [Na]). And condition (C_1) can be weakened, by the same method as in the proof of Theorem 7.1 in [Na], to the following:

($\bar{C}_1$) $\sigma \in C^2$, $\sigma'(v) \ge 0$, but $meas\{v : \sigma'(v) = 0\} = 0$; $v\sigma''(v) > 0$ for all $v \in R\backslash\{0\}$.

When considering the convergence of $(v^{\varepsilon,\tau}, u^{\varepsilon,\tau})$ as τ and ε tend to zero, the following two assumptions are used both in [CLL] and [Na]:

(C_4) $h \in C^1$ and there is no interval in which $h(v)$ is affine;

(C_5) $|(v_0, u_0)(x)| \le N_0$, for a suitable constant N_0. However, there is no rate restriction between τ and ε, and oscillatory initial data are allowed.

Using Theorem 14.1.1 and the boundedness estimate from assumptions ($\bar{C}_1$) and (C_2), we have the following theorem:

Theorem 14.3.1 *If h, σ satisfy assumptions (C_1) and (C_2), then there exists a subsequence $(v^{\varepsilon k}, u^{\varepsilon k})$ of global smooth solutions $(v^\varepsilon, u^\varepsilon)$ for the Cauchy problem (14.3.2) with the bounded initial data (14.1.2), that converges strongly to the equilibrium state functions (v, u), determined uniquely by $h(v)$ and $v_0(x)$ (see (E_1)-(E_2) of Theorem 14.1.1).*

14.3.2.   The System of Isentropic Gas Dynamics in Eulerian Coordinates

The system of isentropic gas dynamics in Eulerian coordinates is described by

$$\begin{cases} \rho_t + (\rho u)_x = 0, \\ (\rho u)_t + (\rho u^2 + P(\rho))_x = 0, \end{cases} \qquad (14.3.3)$$

where $\rho, m = \rho u$ and P are the density, the mass and the pressure, respectively. For the case of a polytropic gas $P(\rho) = \kappa\rho^\gamma, \kappa > 0$, where

γ is the adiabatic exponent, the existence of global weak solutions with L^∞ large initial data for this system with general adiabatic exponent $\gamma > 1$ is given in Chapter 8.

Adding a relaxation term to system (14.3.3), we get the following system:

$$\begin{cases} \rho_t + (\rho u)_x = 0, \\ (\rho u)_t + (\rho u^2 + P(\rho))_x + \frac{\rho u - h(\rho)}{\tau} = 0, \end{cases} \tag{14.3.4}$$

which arises in many physical situations such as the flood flows with friction or the river equations (cf. [CLL, Chern, Wh]).

Consider the viscosity solutions of system (14.3.4):

$$\begin{cases} \rho_t + m_x = \varepsilon \rho_{xx}, \\ m_t + (\frac{m^2}{\rho} + P(\rho))_x + \frac{m - h(\rho)}{\tau} \varepsilon m_{xx}, \end{cases} \tag{14.3.5}$$

with bounded initial data:

$$\begin{cases} (\rho^{\varepsilon,\tau,\delta}, m^{\varepsilon,\tau,\delta})|_{t=0} = (\rho_0^\delta(x), m_0^\delta(x)) \equiv (\rho_0(x) + \delta, \rho_0(x)u_0(x)), \\ 0 \le \rho_0(x), \quad |\frac{m_0(x)}{\rho_0(x)}| \le M_0 < \infty. \end{cases}$$
$$\tag{14.3.6}$$

Applying the invariant region principle, Lattanzio and Marcati obtained the *a priori* L^∞ estimate of solutions of the Cauchy problem (14.3.5), (14.3.6) for $h(\rho) = \rho(1 - \rho)$ (cf. [LM]). They also considered the zero relaxation limit for system (14.3.4) in the domain of the density away from vacuum.

Fortunately, since only L^∞ estimate uniformly in the relaxation time and the dissipation parameter is required in Theorem 14.1.1, we can obtain the convergence of solutions, including the vacuum, of the Cauchy problem (14.3.5), (14.3.6) for more general cases.

Theorem 14.3.2 *Let* $P(\rho) \in C^2(0, \infty), P'(\rho) > 0, 2P'(\rho) + \rho P''(\rho) \ge 0$ *for* $\rho > 0$ *and*

$$\int_c^\infty \frac{\sqrt{P'(\rho)}}{\rho} d\rho = \infty, \quad \int_0^c \frac{\sqrt{P'(\rho)}}{\rho} d\rho < \infty, \quad \forall c > 0.$$

Suppose that there exists a region

$$\Sigma_8 = \{(\rho, m) : w \le N, z \ge -L\},$$

for some N, L, such that the curve $m = h(\rho)$ and the initial data $(\rho_0^\delta(x), m_0^\delta(x))$ are inside the region Σ_8, and $m = h(\rho)$ passes the two intersection points $(\rho, m) = (0,0)$ and $(\rho, m) = (\bar\rho, \bar m), \bar\rho > 0$, of the curves $w = N$ and $z = -L$ (see Figure 14.1).

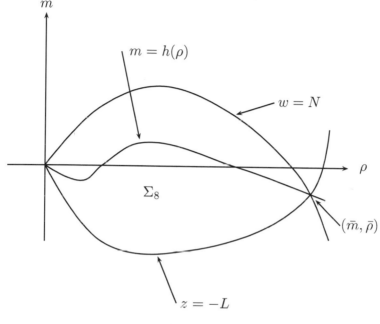

FIGURE 14.1

Then, for any fixed ε, τ and δ, the Cauchy problem (14.3.5), (14.3.6) has a unique smooth solution $(\rho^{\varepsilon,\tau,\delta}, u^{\varepsilon,\tau,\delta})$ satisfying

$$0 < c(t, \varepsilon, \delta) \leq \rho^{\varepsilon,\tau,\delta} \leq M, \quad |u^{\varepsilon,\tau,\delta}| \leq M, \tag{14.3.7}$$

where $c(t, \varepsilon, \delta)$ is a positive constant. Moreover, there exists a subsequence, still denoted $(\rho^{\varepsilon,\tau,\delta}, u^{\varepsilon,\tau,\delta})$, that converges pointwisely almost everywhere:

$$(\rho^{\varepsilon,\tau,\delta}, \rho^{\varepsilon,\tau,\delta} u^{\varepsilon,\tau,\delta}) \to (\rho, m)$$

as $\delta, \varepsilon \downarrow 0^+$ with $\tau = o(\varepsilon)$, where the limit functions (ρ, m) are the equilibrium state functions determined uniquely by $h(\rho)$ and $\rho_0(x)$ (see (E_1)-(E_2) of Theorem 14.1.1).

Proof. To prove the estimates in (14.3.7), we multiply (14.3.5) by (w_ρ, w_m) and (z_ρ, z_m), respectively, to obtain

$$w_t + \lambda_2 w_x + \frac{\rho u - h(\rho)}{\tau \rho} = \varepsilon w_{xx} + \frac{2\varepsilon}{\rho} \rho_x w_x$$

$$- \frac{\varepsilon}{2\rho^2 \sqrt{P'(\rho)}} (2P' + \rho P'') \rho_x^2,$$

$$z_t + \lambda_1 z_x + \frac{\rho u - h(\rho)}{\tau \rho} = \varepsilon z_{xx} + \frac{2\varepsilon}{\rho} \rho_x z_x$$

$$+ \frac{\varepsilon}{2\rho^2 \sqrt{P'(\rho)}} (2P' + \rho P'') \rho_x^2.$$

Then the assumptions on $P(\rho)$ yield

$$\begin{cases} w_t + \lambda_2 w_x + \dfrac{\rho u - h(\rho)}{\tau \rho} \leq \varepsilon w_{xx} + \dfrac{2\varepsilon}{\rho} \rho_x w_x, \\[2mm] z_t + \lambda_1 z_x + \dfrac{\rho u - h(\rho)}{\tau \rho} \geq \varepsilon z_{xx} + \dfrac{2\varepsilon}{\rho} \rho_x z_x. \end{cases} \qquad (14.3.8)$$

If the curve $m = h(\rho)$ passes the two intersection points $(0,0), (\bar{\rho}, \bar{m})$ of curves $w = N, z = -L$ and is above the curve $z = -L$ and below the curve $w = N$ as $0 \leq \rho \leq \bar{\rho}$, then it is easy to check that the region $\Sigma_8 = \{(\rho, m) : w \leq N, z \geq -L\}$ is an invariant region. Thus we obtain the estimates $0 \leq \rho^{\varepsilon,\tau,\delta} \leq M$ and $|u^{\varepsilon,\tau,\delta}| \leq M$ for a suitable constant M, since $\int_c^\infty \frac{\sqrt{P'(\rho)}}{\rho} d\rho = \infty$ and $\int_0^c \frac{\sqrt{P'(\rho)}}{\rho} d\rho < \infty$ for any constant $c > 0$.

The positive lower bound of ρ in (14.3.7) can be obtained by the last part of Theorem 1.0.2. Thus Theorem 14.3.2 follows from Theorem 14.1.1. ∎

14.3.3. Extended Models of Traffic Flows: L^∞ Solutions

Consider the viscosity solutions to the extended model of traffic flows:

$$\begin{cases} \rho_t + (\rho u)_x = \varepsilon \rho_{xx}, \\[2mm] u_t + (\frac{u^2}{2} + g(\rho))_x + \frac{u - h(\rho)}{\tau} = \varepsilon u_{xx}, \end{cases} \qquad (14.3.9)$$

with bounded initial data

$$(\rho, u)|_{t=0} = (\rho_0(x), u_0(x)), \quad \rho_0(x) \geq 0. \qquad (14.3.10)$$

The existence of weak solutions to the corresponding hyperbolic system of (14.3.9),

$$\begin{cases} \rho_t + (\rho u)_x = 0, \\ u_t + (\frac{u^2}{2} + g(\rho))_x = 0, \end{cases} \tag{14.3.11}$$

is obtained in Chapters 9 and 10.

The study of the zero relaxation limit for system (14.3.11) with a singular relaxation term,

$$\begin{cases} \rho_t + (\rho u)_x = 0, \\ u_t + (\frac{u^2}{2} + g(\rho))_x + \frac{u - h(\rho)}{\tau} = 0, \end{cases} \tag{14.3.12}$$

was started by Schochet in [Sc]. System (14.3.12) was derived for car traffic flows (cf. [Wh]) and its existence of classical solutions for all time was obtained in [Sc] for the case $f(\rho) = \frac{\mu}{\tau} \log \rho$, provided that τ is sufficiently small and $\tau \leq \mu^{3+\alpha}$, $\alpha > 0$.

Using Theorem 14.1.1, we have the following theorem:

Theorem 14.3.3 *Let $g'(\rho) > 0$, and $g'(\rho)/\rho$ be a nondecreasing function. Suppose that there exist two constants N, L such that the curve $u = h(\rho)$ passes the unique intersection point $(\bar{\rho}, \bar{u})$ of curves $w = N, z = -L$; the curve $u - h(\rho)$ and the initial data $(\rho_0(x), u_0(x))$ are in the region*

$$\Sigma_9 = \{(\rho, u) : w \leq N, z \geq -L, \rho \geq 0\}$$

as $0 \leq \rho \leq \bar{\rho}$ (see Figure 14.2). Then, for any fixed ε and τ, the Cauchy problem (14.3.9), (14.3.10) has a unique smooth solution $(\rho^{\varepsilon,\tau}, u^{\varepsilon,\tau})$ satisfying

$$0 \leq \rho^{\varepsilon,\tau} \leq M, \quad |u^{\varepsilon,\tau}| \leq M. \tag{14.3.13}$$

Moreover, there exists a subsequence of $(\rho^{\varepsilon,\tau}, u^{\varepsilon,\tau})$ that converges pointwisely to (ρ, u), as ε tend to zero with $\tau = o(\varepsilon)$, where (ρ, u) are the equilibrium state functions, determined uniquely by $h(\rho)$ and $\rho_0(x)$ (see (E_1)-(E_2) of Theorem 14.1.1).

Proof. Noticing the conclusions in Theorem 14.1.1, the crux to prove Theorem 14.3.3 is still the boundedness estimates in (14.3.13).

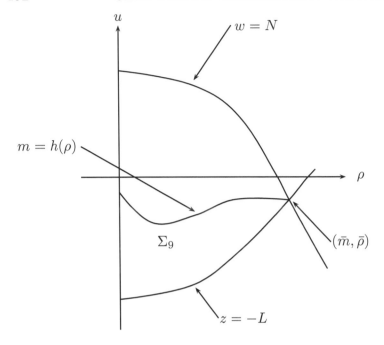

FIGURE 14.2

Multiplying (14.3.9) by (w_ρ, w_u) and (z_ρ, z_u), respectively, where w and z are Riemann invariants of system (14.3.11), we have

$$w_t + \lambda_2 w_x + \tfrac{1}{\tau}(u - h(\rho))\varepsilon w_{xx} - (\sqrt{g'/\rho})'\rho_x^2 \le \varepsilon w_{xx},$$

$$z_t + \lambda_1 z_x + \tfrac{1}{\tau}(u - h(\rho))\varepsilon z_{xx} + (\sqrt{g'/\rho})'\rho_x^2 \ge \varepsilon z_{xx}.$$

From the assumptions, there exist two constants N, L such that the curve $u = h(\rho)$ passes the unique intersection point $(\bar{\rho}, \bar{u})$ of curves $w = N, z = -L$, and the curve $u = h(\rho)$ and the initial data $(\rho_0(x), u_0(x))$ are in the region $\Sigma_9 = \{(\rho, u) : w \le N, z \ge -L, \rho \ge 0\}$ as $0 \le \rho \le \bar{\rho}$ (see Figure 14.2). Then Σ_9 must be an invariant region. This completes the proof of Theorem 14.3.3. ∎

14.4   Proof of Theorem 14.1.2

In this section we prove Theorem 14.1.2 in several steps.

Proof of (I). The local existence and regularity of solutions for $t \in (0, t_0)$ for the Cauchy problem (14.1.1), (14.1.2) can be obtained by applying the Banach contraction mapping theorem to an integral representation of (14.1.1), where the local time depends only on ϵ, τ, and the $L^1 \cap L^\infty$ norm of the initial data $(v_0 - \bar{v}, u_0 - \bar{u})$. The global existence is based on the local existence and the global *a priori* L^∞ estimate (14.1.7) proved below. Set

$$\tilde{p}(v, u) = \bar{p}(v, u) - \bar{p}(\bar{v}, \bar{u}) - \bar{p}_v(\bar{v}, \bar{u})(v - \bar{v}) - \bar{p}_u(\bar{v}, \bar{u})(u - \bar{u}),$$

and similarly $\tilde{\eta}(v, u)$ and $\tilde{p}(v, u)$.

Since $\tilde{\bar{p}}(v, u)$ is strictly convex, then

$$\tilde{\bar{p}}_{vv}(v, u)v_x^2 + 2\tilde{\bar{p}}_{vu}(v, u)v_x u_x + \tilde{\bar{p}}_{uu}(v, u)u_x^2 \geq C_2(v_x^2 + u_x^2), \quad (14.4.1)$$

for some constant $C_2 > 0$.

Since $\tilde{\eta}(v, u)$ is an entropy of (14.1.3), we set the corresponding entropy flux by $Q(v, u)$. Observing $\bar{p}(v, u) = \eta(v, u) + p(v, u)$, we multiply system (14.1.1) by $(\tilde{\bar{p}}_v, \tilde{\bar{p}}_u)$ to obtain

$$\tilde{\bar{p}}_t(v, u) + Q_x(v, u) + \tilde{p}_v f(v, u)_x + \tilde{p}_u g(v, u)_x$$
$$+ \frac{1}{\tau} p_1(v, u)a(v, u)(u - h(v))^2 + \epsilon C_2(v_x^2 + u_x^2) \quad (14.4.2)$$
$$\leq \epsilon \tilde{\bar{p}}_{xx}(v, u).$$

Noticing (14.2.4)-(14.2.5) and the conditions of (B_1)-(B_4), we have

$$\tilde{\bar{p}}_t(v, u) + \bar{Q}_x(v, u) + \frac{1}{2\tau} c_0 a(v, u)(u - h(v))^2$$
$$+ (\epsilon C_2 - C_3 \tau)(v_x^2 + u_x^2) \quad (14.4.3)$$
$$\leq \epsilon \tilde{\bar{p}}_{xx}(v, u),$$

for a suitable function $\bar{Q}(v, u)$ and a constant C_3 depending on the bounds of second derivatives of p. Therefore, we have the following estimates:

$$\begin{cases} \|(v(\cdot, t) - \bar{v}, u(\cdot, t) - \bar{u})\|_{L^2(R)} + \|(\epsilon v_x^2, \epsilon u_x^2)\|_{L^1(R_+^2)} \leq M, \\ \left\| \dfrac{a(v, u)(u - h(v))^2}{\tau} \right\|_{L^1(R_+^2)} \leq M, \end{cases}$$
$$(14.4.4)$$

provided that $\epsilon C_2 > 2C_3 \tau$.

Multiplying the first equation in (14.1.1) by v_{xx} and the second by u_{xx}, adding the outcome, and then integrating by parts on R_+^2, we have

$$
\frac{1}{2}\int_{-\infty}^{\infty}(v_x^2+u_x^2)dx + \varepsilon\int_0^t\int_{-\infty}^{\infty}(v_{xx}^2+u_{xx}^2)dxdt
$$
$$
= \frac{1}{2}\int_{-\infty}^{\infty}(v_x^2+u_x^2)(x,0)dx + \int_0^t\int_{-\infty}^{\infty}\Big\{f(v,u)_x v_{xx} \qquad (14.4.5)
$$
$$
+[g(v,u)_x+\tfrac{1}{\tau}\alpha(v,u)(u-h(v))]u_{xx}\Big\}dxdt.
$$

Then it follows from (14.4.5), the estimates in (14.4.4) and the growth conditions (B_1)-(B_2) that

$$
\int_{-\infty}^{\infty}(v_x^2+u_x^2)dx \;\leq\; c_0(\varepsilon,\tau)(1+|v|_\infty^{2(q-1)}+|u|_\infty^{2(q-1)}+|\alpha(v,u)|_\infty)
$$
$$
\leq c_0(\varepsilon,\tau)(1+|v|_\infty^{2(q-1)}+|u|_\infty^{2(q-1)}+|v|_\infty^r+|u|_\infty^r).
$$
$$
(14.4.6)
$$

Since

$$
v^2+u^2 \;=\; \int_{-\infty}^{x}(v^2+u^2)_x dx \leq 2|v|_2|v_x|_2+2|u|_2|u_x|_2
$$
$$
\leq c_1(\varepsilon,\tau)(1+|v|_\infty^{2(q-1)}+|u|_\infty^{2(q-1)}+|v|_\infty^r+|u|_\infty^r)^{\frac{1}{2}},
$$
$$
(14.4.7)
$$

where $2(q-1)<4$ and $r<4$, then

$$
|(v,u)(x,t)| \leq C(\varepsilon,\tau), \quad t>0,
$$

for some constant $C(\varepsilon,\tau)>0$. Thus (I) is proved.

Proof of (II). Multiplying the first equation in (14.1.1) by $p|v-\bar v|^{p-2}(v-\bar v)$, we have

$$
(|v-\bar v|^p)_t + P(v,u)_x - p(p-1)(f(v,u)-f(v,h(v)))|v-\bar v|^{p-2}v_x
$$
$$
= \varepsilon(|v-\bar v|^p)_{xx} - \varepsilon p(p-1)|v-\bar v|^{p-2}v_x^2,
$$
$$
(14.4.8)
$$

where $P(v,u)=p|v-\bar v|^{p-2}(v-\bar v)(f(v,u)-f(v,h(v)))+\int^v p|v-\bar v|^{p-2}(v-\bar v)f'(v,h(v))dv$.

Since $|f(v, u) - f(v, h(v))| = |f_u(v, \mu)(u - h(v))|$, where μ takes a value between u and $h(v)$, it follows from (14.4.8) that

$$(|v - \bar{v}|^p)_t + P(v, u)_x + \varepsilon p(p - 1)|v - \bar{v}|^{p-2}v_x^2$$

$$\leq \varepsilon(|v - \bar{v}|^p)_{xx} + \frac{\alpha(v, u)(u - h(v))^2}{\tau}$$

$$+ \frac{\tau p(p - 1)|v - \bar{v}|^{p-2}|f_u(v, \mu)|^2}{\alpha(v, u)}p(p - 1)|v - \bar{v}|^{p-2}v_x^2$$

$$\leq \varepsilon(|v - \bar{v}|^p)_{xx} + \frac{\alpha(v, u)(u - h(v))^2}{\tau} + \frac{\varepsilon}{2}p(p - 1)|v - \bar{v}|^{p-2}v_x^2,$$

$$(14.4.9)$$

provided that $2\tau p(p - 1)M \leq \varepsilon$, where

$$\frac{p(p - 1)|v - \bar{v}|^{p-2}|f_u(v, \mu)|^2}{\alpha(v, u)} \leq M$$

from the condition $2k(q - 1) + p - 2 \leq r$. Integrating (14.4.9) by parts on $R \times [0, t]$, we have the estimate $\|v - \bar{v}\|_p \leq M$. Thus (II) is proved.

Proof of (III). Since $|f(v, h(v))| \leq c_1 + c_2(|v|^q + |h(v)|^q) \leq M(1 + |v|^q + |v|^{qk})$ and $p > max\{1, q, qk\}$, then (F_1) and (F_2) in (III) can be proved directly by (14.1.8) and the estimate $\|v - \bar{v}\|_p \leq M$.

We can use a similar method as in the proof of Lemma 3.2.4 to prove (14.1.8). This completes the proof of Theorem 14.1.2. ∎

Remark 14.4.1 *For the simplicity of proof, in Chapter 3, we need the technical condition (3.1.6) and the growth condition*

$$|f(v)| \leq c_1 + c_2|v|^q, \quad 2q = p$$

on the nonlinear flux function $f(v)$ to prove the strong convergence of the viscosity solutions in L^p space. In Theorem 14.1.2, the condition (3.1.6) is removed and the growth condition on nonlinear function $f(v, h(v))$ is weakened to

$$|f(v, h(v))| \leq c_1 + c_2|v|^q, \quad q < p.$$

The details can be found in [Sz, Lu9].

14.5 Applications of Theorem 14.1.2

In this section we are concerned with the zero relaxation and dissipation limits for some physical models, without bounded L^∞ estimates, such as the system of isentropic gas dynamics in Lagrangian coordinates, the system of elasticity in the case of $\sigma''(v) \cdot v < 0$, for all $v \in R\backslash\{0\}$ and the models of traffic flows with stiff relaxation terms.

14.5.1. System of Isentropic Gas Dynamics in Lagrangian Coordinates

Consider the viscosity solutions of the system of isentropic gas dynamics with relaxation terms in Lagrangian coordinates:

$$\begin{cases} v_t - u_x = \varepsilon v_{xx}, \\ u_t + g(v)_x + \dfrac{1}{\tau}\alpha(v,u)(u - h(v)) = \varepsilon u_{xx}, \end{cases} \tag{14.5.1}$$

with initial data

$$(v,u)|_{t=0} = (v_0(x), u_0(x)), \tag{14.5.2}$$

where $g'(v) \le 0, g''(v) > 0$ as $v > 0$. The two eigenvalues of the corresponding hyperbolic system of (14.5.1) are

$$\lambda_1 = -\sqrt{-g'(v)}, \quad \lambda_2 = \sqrt{-g'(v)},$$

and the corresponding two Riemann invariants are

$$z = u + \int_{v_1}^v \sqrt{-g'(v)}\,dv, \quad w = u - \int_{v_1}^v \sqrt{-g'(v)}\,dv,$$

for a constant $v_1 > 0$.

Let $(v_0(x), u_0(x))$ be in the open region $\Sigma_{10} = \{(v,u) : z > 0, w < 0\}$ and $(v_0(x) - \bar{v}, u_0(x) - \bar{u}) \in L^1 \cap L^p$, where $\bar{v} \ge v_1, \bar{u} = h(\bar{v})$.

Theorem 14.5.1 *(I) If*

$$-\int_{v_1}^v \sqrt{-g'(v)}\,dv \le h(v) \le \int_{v_1}^v \sqrt{-g'(v)}\,dv,$$

$$\left(h'(v)^2 + g'(v)^2 + |h''(v)| \int_{v_0}^{v} \sqrt{-g'(v)}dv\right)/\alpha(v,u) \quad \text{is bounded,}$$

and

$$0 < \alpha_0 \leq \alpha(v,u) \leq c_1(1 + |v|^r + |u|^r), \quad r \in [0,4), \quad \text{as } v \geq v_1,$$

then, for fixed ε and τ satisfying $M_1\tau \leq \varepsilon$ for a suitable large constant M_1, the Cauchy problem (14.5.1)-(14.5.2) has a unique smooth solution $(v^{\varepsilon,\tau}, u^{\varepsilon,\tau})$ satisfying

$$v_1 \leq v^{\varepsilon,\tau} \leq C(\varepsilon,\tau), \quad |u^{\varepsilon,\tau}| \leq C(\varepsilon,\tau), \tag{14.5.3}$$

for some positive constant $C(\varepsilon,\tau)$ depending on ε and τ.

(II) Let the conditions of (I) be satisfied and $|v - \bar{v}|^{p-2}/\alpha(v,u)$ be finite as $v \geq v_1$. Then $\|v^{\varepsilon,\tau} - \bar{v}\|_p \leq M$.

(III) Let the conditions of (I) and (II) be satisfied, $|h(v)| \leq c_3(1 + |v|^k)$ and $p > k$. Then there exists a subsequence of $(v^{\varepsilon,\tau}, u^{\varepsilon,\tau})$ converging pointwisely almost everywhere to (v, u), as $\epsilon \to 0$ with $\tau = o(\varepsilon)$, where (v, u) are the equilibrium state functions, determined uniquely by $h(v)$ and $v_0(x)$ (see (F_1)-(F_2) of Theorem 14.1.2).

Proof. We only give the proof of some estimates similar to those in (14.4.4). The remaining can be completed similarly as in the proof of Theorem 14.1.2.

Multiplying (14.5.1) by (w_v, w_u) and (z_v, z_u), respectively, we have

$$\begin{cases} w_t + \lambda_2 w_x + \frac{1}{\tau}\alpha(v,u)(u - h(v)) = \varepsilon w_{xx} + \varepsilon(\sqrt{-g'(v)})'v_x^2 \leq \varepsilon w_{xx}, \\ \\ z_t + \lambda_1 z_x + \frac{1}{\tau}\alpha(v,u)(u - h(v)) = \varepsilon z_{xx} - \varepsilon(\sqrt{-g'(v)})'v_x^2 \leq \varepsilon z_{xx}. \end{cases} \tag{14.5.4}$$

If $-\int_{v_1}^{v} \sqrt{-g'(v)}dv \leq h(v) \leq \int_{v_1}^{v} \sqrt{-g'(v)}dv$ as $v \geq v_1$, then the curve $u = h(v)$ is inside the region Σ_{10} (see Figure 14.3).

Thus we can get directly from inequalities (14.5.4) that Σ_{10} is an invariant region. From this we have the estimates

$$v \geq v_1, \quad |u| \leq \int_{v_1}^{v} \sqrt{-g'(v)}dv. \tag{14.5.5}$$

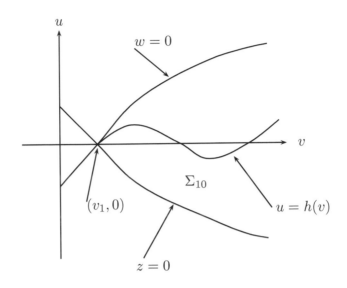

FIGURE 14.3

Let

$$p(v, u) = \frac{u^2}{2} - h(v)u$$
$$+4 \int_{v_1}^{v} \int_{v_1}^{s} \left(|h''(m)| \int_{v_1}^{m} \sqrt{-g'(n)} dn + h'(m)^2 + 1 \right) dm\, ds,$$

$$(14.5.6)$$

and

$$\bar{p}(v, u) = p(v, u) - p(\bar{v}, \bar{u}) - p_v(\bar{v}, \bar{u})(v - \bar{v}) - p_u(\bar{v}, \bar{u})(u - \bar{u}).$$
$$(14.5.7)$$

Multiplying system (14.5.1) by $(\bar{p}_v, \bar{p}_u)$, we have from (14.2.4)-(14.2.5) that

$$\bar{p}_t(v, u) + \bar{q}_x(v, u) + (-h'(v)u_x + \bar{p}_{vv}(v, u)v_x)(u - h(v))$$
$$+ h'(v)^2(u - h(v))v_x + g'(v)(u - h(v))v_x + \tfrac{1}{\tau}\alpha(v, u)(u - h(v))^2$$
$$= \varepsilon \bar{p}_{xx}(v, u) - \varepsilon(\bar{p}_{vv}(v, u)v_x^2 + 2\bar{p}_{vu}(v, u)v_x u_x + \bar{p}_{uu}(v, u)u_x^2),$$
$$(14.5.8)$$

where

$$
\begin{aligned}
\bar{q}(v, u) &= -\bar{p}_v(v, u)(u - h(v)) - \int_{v_1}^{v} \bar{p}_v(s, h(s))h'(s)ds \\
&\quad + \int_{v_1}^{v} \bar{p}_u(s, h(s))g'(s)ds.
\end{aligned}
$$

It follows from (14.5.6)-(14.5.7) and the second estimate in (14.5.5) that

$$
\begin{aligned}
&\bar{p}_{vv}(v, u)v_x^2 + 2\bar{p}_{vu}(v, u)v_x u_x + \bar{p}_{uu}(v, u)u_x^2 \\
&\geq C_2\{(|h''(v)| \int_{v_1}^{v} \sqrt{-g'(s)}ds + h'(v)^2 + 1)v_x^2 + u_x^2\},
\end{aligned} \tag{14.5.9}
$$

for some constant $C_2 > 0$. Thus it follows from (14.5.8) and the conditions in (I) that

$$
\begin{aligned}
&\bar{p}_t(v, u) + q_x(v, u) + \tfrac{1}{2\tau}\alpha(v, u)(u - h(v))^2 \\
&\quad + (\varepsilon C_2 - M\tau)\Big\{(|h''(v)| \int_{v_1}^{v} \sqrt{-g'(s)}ds + h'(v)^2 + 1)v_x^2 + u_x^2\Big\} \\
&\leq \varepsilon \bar{p}_{xx}(v, u),
\end{aligned} \tag{14.5.10}
$$

which yields that

$$
\left\{
\begin{aligned}
&\|(v(\cdot, t) - \bar{v}, u(\cdot, t) - \bar{u})\|_{L^2(R)} + \|\frac{\alpha(v, u)(u - h(v))^2}{\tau}\|_{L^1(R_+^2)} \leq M; \\
&\|\varepsilon\{(|h''(v)| \int_{v_1}^{v} \sqrt{-g'(s)}ds + h'(v)^2 + 1)v_x^2 + u_x^2\}\|_{L^1(R_+^2)} \leq M,
\end{aligned}
\right. \tag{14.5.11}
$$

provided that $2M\tau \leq \varepsilon C_2$. Therefore, the proof can be similarly completed as in the proof of Theorem 14.1.2. ∎

14.5.2. Models of Traffic Flows: L^p Solutions

We consider the viscosity solutions of the models of traffic flows with relaxation terms

$$
\left\{
\begin{aligned}
&\rho_t + (\rho u)_x = \varepsilon \rho_{xx}, \\
&u_t + (\tfrac{u^2}{2} + g(\rho))_x + \tfrac{1}{\tau}\alpha(\rho, u)(u - h(\rho)) = \varepsilon u_{xx},
\end{aligned}
\right. \tag{14.5.12}
$$

with initial data

$$(\rho, u)|_{t=0} = (\rho_0, u_0), \qquad (14.5.13)$$

where (ρ_0, u_0) satisfies

$$(\rho_0(x) - \bar{\rho}, u_0(x) - \bar{u}) \in L^1 \cap L^p, \quad 1 \le p < \infty, \qquad (14.5.14)$$

with constants $(\bar{\rho}, \bar{u}), \bar{\rho} > 0$, and $\bar{u} = h(\bar{\rho})$.

For the model of traffic flow, $g(\rho) = \log \rho$. If $g'(\rho)/\rho$ is a nonincreasing function, system (14.5.12) generally does not have an *a priori* L^∞ estimate. We study the compactness of solutions $(\rho^{\varepsilon,\tau}, u^{\varepsilon,\tau})$ of the Cauchy problem (14.5.12), (14.5.13) in L^p.

Two eigenvalues of the corresponding hyperbolic system (14.3.11) of (14.5.12) are

$$\lambda_1 = u - \sqrt{\rho g'(\rho)}, \quad \lambda_2 = u + \sqrt{\rho g'(\rho)}$$

and the corresponding two Riemann invariants are

$$z = u - \int_{\rho_1}^{\rho} \sqrt{g'(s)/s}\, ds, \quad w = u + \int_{\rho_1}^{\rho} \sqrt{g'(s)/s}\, ds,$$

where $0 < \rho_1 \le \bar{\rho}$ is a constant.

Let

(D_1) $g'(\rho) > 0$, $(g'(\rho)/\rho)' \le 0$, as $\rho \ge \rho_1$;

(D_2) $0 < \alpha_0 \le \alpha(\rho, u) \le c_1 + c_2(|\rho|^r + |u|^r)$, $0 \le r < 4$;

(D_3)
$$\frac{(\rho h'(\rho))^2 + h(\rho)^2 + g'(\rho)^2 + \rho^2 + u^2 + \rho^2 |h''(\rho)| \int_{\rho_1}^{\rho} \sqrt{g'(s)/s}\, ds}{\alpha(\rho, u)}$$

is bounded when $|u| \le \int_{\rho_1}^{\rho} \sqrt{g'(s)/s}\, ds$;

(D_4) $\rho^p / \alpha(\rho, u) \le M$;

(D_5) $|h(\rho)| \le c_3 + c_4 |\rho|^k$.

Theorem 14.5.2 *(I) Let conditions (D_1)-(D_3) be satisfied, and the initial data (14.5.13) be in the region $\Sigma = \{(\rho, u) : w \ge 0, z \le 0\}$. Suppose*

$$-\int_{\rho_1}^{\rho} \sqrt{g'(s)/s}\, ds \le h(\rho) \le \int_{\rho_1}^{\rho} \sqrt{g'(s)/s}\, ds,$$

as $\rho \geq \rho_1$. Then, for fixed ε, τ satisfying $M_1 \tau \leq \varepsilon$ for a suitable large constant M_1, the Cauchy problem (14.5.12), (14.5.13) has a unique smooth solution $(\rho^{\varepsilon,\tau}, u^{\varepsilon,\tau})$ satisfying

$$\rho_1 \leq \rho^{\varepsilon,\tau} \leq C(\varepsilon,\tau), \quad |u^{\varepsilon,\tau}| \leq C(\varepsilon,\tau), \tag{14.5.15}$$

for some constant $C(\varepsilon, \tau)$ depending on ε, τ.

(II) Let the conditions of (I) and (D_4) be satisfied. Then $\|\rho^{\varepsilon,\tau} - \bar{\rho}\|_p \leq M$.

(III) Let the conditions of (I), (II) and (D_5) be satisfied and $p > 1 + k$. Then there exists a subsequence of $(\rho^{\varepsilon,\tau}, u^{\varepsilon,\tau})$ converging pointwisely almost everywhere to (ρ, u), as $\epsilon \to 0$ with $\tau = o(\varepsilon)$, where the limit functions (ρ, u) are the equilibrium states, determined uniquely by $h(\rho)$ and $\rho_0(x)$ (see (F_1)-(F_2) of Theorem 14.1.2).

Proof. Similar to the proof of Theorem 14.5.1, we can prove that Σ is an invariant region. Hence we have the following estimates:

$$\rho \geq \rho_1, \quad |u| \leq \int_{\rho_1}^{\rho} \sqrt{g'(s)/s}\, ds. \tag{14.5.16}$$

Choose

$$p(\rho, u) = \frac{u^2}{2} - h(\rho)u$$

$$+ 4 \int_{\rho_1}^{\rho} \int_{\rho_1}^{m} (|h''(m)|\sqrt{g'(n)/n}\, dn + (h')^2(m) + 1) dm\, ds,$$

and let

$$\bar{p}(\rho, u) = p(\rho, u) - p(\bar{\rho}, \bar{u}) - p_\rho(\bar{\rho}, \bar{u})(\rho - \bar{\rho}) - p_u(\bar{\rho}, \bar{u})(u - \bar{u}).$$

Multiplying system (14.5.12) by $(\bar{p}_\rho, \bar{p}_u)$, we can obtain the following estimates from (D_3) and the second estimate in (14.5.16):

$$\begin{cases} \|(\rho(\cdot, t) - \bar{\rho}, u(\cdot, t) - \bar{u})\|_{L^2(R)} + \|\dfrac{\alpha(\rho, u)(u - h(\rho))^2}{\tau}\|_{L^1(R_+^2)} \leq M; \\[4mm] \|\varepsilon\{(|h''(\rho)| \displaystyle\int_{\rho_1}^{\rho} \sqrt{g'(s)/s}\, ds + h'(\rho)^2 + 1)\rho_x^2 + u_x^2\}\|_{L^1(R_+^2)} \leq M. \end{cases}$$

$$\tag{14.5.17}$$

Using (D_1)-(D_2) and the estimates in (14.5.17), we can get the estimates in (14.5.15). Then (I) is proved.

Since $f_u = \rho$, the proof of (II) follows from (14.4.8) and (D_4).

Since $\rho^2/\alpha(\rho, u)$ is bounded, similar to the proof of Theorem 14.1.2, we can complete the proof of (III). This completes the proof of Theorem 14.5.2. ■

14.6   Related Results

The nonlinear stability of weak, smooth travelling waves and rarefaction waves for hyperbolic systems of conservation laws with relaxation were first analyzed by T.-P. Liu (cf. [Liu]), in which it is indicated that the effect of relaxation is closely related to a viscous effect when the solution is near a constant equilibrium state (also see [Chern]). By introducing the theory of the compensated compactness, a simple model of combustion (3.3.3) with infinite reaction rate k (whose reciprocal is related to the zero relaxation) was first studied in [Lu7]. Later, the zero relaxation limit of solutions with large oscillation, for hyperbolic conservation laws with relaxation terms containing both damping and sink mechanisms, was systematically studied in Chen-Liu [ChL] and Chen-Levermore-Liu [CLL] also by the compensated compactness method. All the results in this chapter are from [Lu9].

Chapter 15

Hyperbolic Systems with Stiff Relaxation

In this chapter, we are concerned with singular limits of relaxation approximated solutions (v^τ, u^τ) to the Cauchy problem of general 2×2 system of quasilinear conservation laws over a one-dimensional spatial domain in the form

$$\begin{cases} v_t + f(v, u)_x = 0, \\ u_t + g(v, u)_x + \frac{1}{\tau}(u - h(v)) = 0, \end{cases} \tag{15.0.1}$$

with initial data

$$(v, u)|_{t=0} = (v_0(x), u_0(x)). \tag{15.0.2}$$

The eigenvalues of system (15.0.1) satisfy the following characteristic equation:

$$\lambda^2 - (f_v + g_u)\lambda + f_v g_u - g_v f_u = 0. \tag{15.0.3}$$

System (15.0.1) is assumed to be hyperbolic (it is not necessary to be strictly hyperbolic), that is, the two eigenvalues or characteristic speeds

$$\lambda_1 = \frac{1}{2}\left(f_v + g_u - \sqrt{(f_v - g_u)^2 + 4g_v f_u}\right) \tag{15.0.4}$$

and

$$\lambda_2 = \frac{1}{2}\left(f_v + g_u + \sqrt{(f_v - g_u)^2 + 4g_v f_u}\right) \tag{15.0.5}$$

are real.

Formally, when the relaxation time $\tau \to 0$, we have from the second equation in (15.0.1) that $u = h(v)$, which implies, from the first equation in (15.0.1), the local equilibrium equation just as the following scalar conservation law:

$$v_t + f(v, h(v))_x = 0. \tag{15.0.6}$$

In the Chapman-Enskog expansion one seeks to identify the effective response of the relaxation process as it approaches the line of local equilibria $u = h(v)$. It is postulated that the relaxing variable u^τ can be described in an asymptotic expansion that involves only the local macroscopic values v^τ and its derivatives, i.e.,

$$u^\tau = h(v^\tau) + \tau s^\tau(v^\tau, u^\tau, v_x^\tau, u_x^\tau, \cdots) + O(\tau^2). \tag{15.0.7}$$

To calculate the form of s^τ, we use system (15.0.1) to get (for simplicity, we omit the superscript τ)

$$\begin{aligned}
0 &= v_t + f(v, u)_x \\
&= v_t + f(v, h(v) + \tau s + O(\tau^2))_x \\
&= v_t + f(v, h(v))_x + \tau(s f_u(v, h(v)))_x + O(\tau^2)
\end{aligned} \tag{15.0.8}$$

and

$$\begin{aligned}
0 &= u_t + g(v, u)_x + \frac{1}{\tau}(u - h(v)) \\
&= h(v)_t + g(v, h(v))_x + s + O(\tau) \\
&= h'(v)v_t + \left(\frac{dg(v, h(v))}{dv}\right)v_x + s + O(\tau).
\end{aligned} \tag{15.0.9}$$

Use (15.0.8) to eliminate v_t in (15.0.9) and obtain

$$\begin{aligned}
-s &= h'(v)(-f(v, h(v))_x + \left(\frac{dg(v, h(v))}{dv}\right)v_x + O(\tau) \\
&= \left(\frac{dg(v, h(v))}{dv} - h'(v)\frac{df(v, h(v))}{dv}\right)v_x + O(\tau).
\end{aligned} \tag{15.0.10}$$

From (15.0.8) and (15.0.10), on $u = h(v)$, we have

$$\begin{aligned}
v_t + f(v, h(v))_x &= -\tau(s f_u(v, h(v)))_x + O(\tau^2) \\
&= \tau\left(\left(\frac{dg(v, h(v))}{dv} - h'(v)\frac{df(v, h(v))}{dv}\right)f_u(v, h(v))v_x\right)_x \\
&= \tau\left(f_u(g_v + (g_u - f_v)h'(v) - f_u(h'(v))^2)v_x\right)_x.
\end{aligned} \tag{15.0.11}$$

Let

$$\phi(v) = f_u(g_v + (g_u - f_v)h'(v) - f_u(h'(v))^2) \text{ on } u = h(v).$$

Then it is easy to prove that

$$\phi(v) = (\lambda_2(v, h(v)) - \lambda(v))(\lambda(v) - \lambda_1(v, h(v))), \qquad (15.0.12)$$

where $\lambda(v) = \frac{df(v,h(v))}{dv}$ is the eigenvalue of the equilibrium equation (15.0.6).

In fact, on $u = h(v)$, we have

$$
\begin{aligned}
&(\lambda_2(v, h(v)) - \lambda(v))(\lambda(v) - \lambda_1(v, h(v))) \\
&= \left(\tfrac{1}{2}(f_v + g_u + \sqrt{(f_v - g_u)^2 + 4g_v f_u}) - f_v - f_u h'(v) \right) \\
&\quad \times \left(f_v + f_u h'(v) - \tfrac{1}{2}(f_v + g_u - \sqrt{(f_v - g_u)^2 + 4g_v f_u}) \right) \\
&= \left(\tfrac{1}{2}\sqrt{(f_v - g_u)^2 + 4g_v f_u} - (\tfrac{1}{2}(f_v - g_u) + f_u h'(v)) \right) \\
&\quad \times \left(\tfrac{1}{2}\sqrt{(f_v - g_u)^2 + 4g_v f_u} + (\tfrac{1}{2}(f_v - g_u) + f_u h'(v)) \right) \\
&= f_u(g_v + (g_u - f_v)h'(v) - f_u(h'(v))^2).
\end{aligned}
$$

$$(15.0.13)$$

Therefore, Equation (15.0.11) is a stable degenerate parabolic equation if the following subcharacteristic condition is satisfied:

$$\lambda_1(v, h(v)) \le \lambda(v) \le \lambda_2(v, h(v)). \qquad (15.0.14)$$

Now we give the definition of entropy-entropy flux pair $(\eta(v, u), q(v, u))$ for the relaxation system (15.0.1) as follows:

Definition 15.0.1 *A pair of functions $(\eta(v, u), q(v, u))$ is called an entropy-entropy flux pair of (15.0.1) if*

(e_1) $(q_v, q_u)(f_v \eta_v + g_v \eta_u, f_u \eta_v + g_u \eta_u),$

(e_2) $\eta_u(v, h(v)) = 0.$

An entropy $\eta(v, u)$ is called convex if

(e_3) $\eta_{vv} a^2 + 2\eta_{vu} ab + \eta_{uu} b^2 \ge c(v, u)(a^2 + b^2)$ *for any vector* $(a, b) \ne (0, 0)$ *in* R^2 *and a nonnegative function* $c(v, u).$

(e_4) *If $c(v, u) \geq c_0 > 0$, for a constant c_0, the entropy is said to be strictly convex.*

Use the entropy-entropy flux equations in (e_1) to eliminate q, and obtain the following entropy equation of system (15.0.1):

$$g_v \eta_{uu} - f_u \eta_{vv} + (f_v - g_u)\eta_{vu} = 0. \tag{15.0.15}$$

If f_u, g_v and $f_v - g_u$ have a common zero factor $Z(v, u)$, then we delete this factor from (15.0.15) and consider entropies of system (15.0.1) to be solutions of the following equation:

$$\frac{g_v}{Z(v, u)}\eta_{uu} - \frac{f_u}{Z(v, u)}\eta_{vv} + \frac{f_v - g_u}{Z(v, u)}\eta_{vu} = 0. \tag{15.0.16}$$

For instance, consider the relaxation problem for the extended model of traffic flows:

$$\begin{cases} \rho_t + (\rho u)_x = 0, \\ u_t + (\frac{u^2}{2} + p(\rho))_x + \frac{u - h(\rho)}{\tau} = 0, \end{cases} \tag{15.0.17}$$

which is nonstrictly hyperbolic on the line $\rho = 0$. One common zero fact of $f_u(\rho, u) = \rho$, $g_\rho(\rho, u) = p'(\rho)$ and $f_\rho(\rho, u) - g_u(\rho, u) = 0$ is ρ.

In Section 15.1, we shall prove a general compactness framework to the singular limits of the general 2×2 system (15.0.1) with the stiff relaxation term, and in Section 15.2, an application of this compactness framework on the nonstrictly hyperbolic system of extended traffic flow (15.0.17) is obtained.

15.1   Relaxation Limits for 2×2 Systems

Before we introduce the compactness framework to the singular limits of the general 2×2 relaxation system (15.0.1) (Theorem 15.1.3), we first prove two basic lemmas.

Lemma 15.1.1 *Let (η, q) be a strictly convex entropy-entropy flux pair of (15.0.1) as defined by (e_1) – (e_4) above. Then the local equilibrium equation (15.0.6) has the strictly convex entropy $l(v) = \eta(v, h(v))$ with the corresponding entropy flux $L(v) = q(v, h(v))$.*

Proof. The function $l(v)$ is clearly an entropy of (15.0.6) since it is a scalar equation. By simple calculations,

$$l''(v) = \eta_{vv}(v, h(v)) + 2\eta_{vu}(v, h(v))h'(v) + \eta_{uu}(v, h(v))(h'(v))^2$$
$$\geq c_0(1 + (h'(v))^2) \geq c_0 > 0,$$

$$(15.1.1)$$

and hence $l(v)$ is strictly convex.

To prove $L(v) = q(v, h(v))$ to be the corresponding entropy flux, we use the equations in (e_1) and the condition in (e_2) to obtain

$$q_v(v, h(v)) = \eta_v(v, h(v))f_v(v, h(v)),$$
$$q_u(v, h(v)) = \eta_v(v, h(v))f_u(v, h(v)).$$

$$(15.1.2)$$

Therefore

$$L'(v) = q_v(v, h(v)) + q_u(v, h(v))h'(v)$$
$$= \eta_v(v, h(v))f_v(v, h(v)) + \eta_v(v, h(v))f_u(v, h(v))h'(v)$$
$$= \eta_v(v, h(v))\frac{df(v, h(v))}{dv},$$

$$(15.1.3)$$

and hence $L(v)$ is the entropy flux corresponding to $l(v)$. ■

Lemma 15.1.2 *Let $l(v)$ be a strictly convex entropy for the local equilibrium equation (15.0.6). Assume that the stability criterion (15.0.14) holds on $u = h(v)$. If*

$$\frac{(\lambda_2(v, h(v)) - \lambda(v))(\lambda(v) - \lambda_1(v, h(v)))}{Z(v, u)^2} > 0, \quad \frac{f_u(v, u)}{Z(v, u)} \neq 0, \quad (15.1.4)$$

where $Z(v, u)$ is a zero factor as given in (15.0.16), then there exists a strictly convex entropy $\eta(v, u)$ for system (15.0.1) over an open set $D_l \subset R^2$ containing the local equilibria curve $u = h(v)$, along which it satisfies $\eta(v, h(v)) = l(v)$.

Proof. If $\eta(v, u)$ is a strictly convex entropy of system (15.0.1), then it must satisfy

(E_1) $\dfrac{g_v}{Z(v,u)}\eta_{uu} - \dfrac{f_u}{Z(v,u)}\eta_{vv} + \dfrac{f_v - g_u}{Z(v,u)}\eta_{vu} = 0,$

(E_2) $\eta_u(v,u)(u - h(v)) \geq 0,$

(E_3) $\eta_{uu} \geq c_1,$ $\eta_{vv}\eta_{uu} - (\eta_{vu})^2 \geq c_2,$ for two positive constants c_1, c_2.

The characteristic curve $u = e(v)$ of Equation (E_1) satisfies the following characteristic equation:

$$\frac{g_v}{Z(v,u)} + \frac{f_v - g_u}{Z(v,u)}e'(v) - \frac{f_u}{Z(v,u)}(e'(v))^2 = 0. \qquad (15.1.5)$$

Therefore, if the conditions (15.1.5) in Lemma 15.1.2 are satisfied, then

$$\begin{aligned}
&\frac{g_v}{Z(v,u)} + \frac{f_v - g_u}{Z(v,u)}h'(v) - \frac{f_u}{Z(v,u)}(h'(v))^2 \\
&= \frac{(\lambda_2(v,h(v)) - \lambda(v))(\lambda(v) - \lambda_1(v,h(v)))}{Z(v,u)^2} \cdot \frac{Z(v,u)}{f_u(v,u)} \neq 0,
\end{aligned}$$

$$(15.1.6)$$

and hence, the curve $u = h(v)$ is not a characteristic curve. Thus the classical Cauchy-Kowalewsky local existence theory ensures that the Cauchy problem for the second-order linear hyperbolic equation (E_1) with the following initial data:

$$\eta(v,h(v)) = l(v), \quad \eta_u(v,h(v)) = 0 \qquad (15.1.7)$$

has a local solution $\eta(v,u)$ over an open domain D_l containing the initial curve or the local equilibria curve $u = h(v)$.

(E_2) can be easily obtained by (E_3). In fact, if (E_3) is true, then

$$\eta_u(v,u)(u - h(v)) = \eta_{uu}(v,\alpha)(u - h(v))^2 \geq 0,$$

where $\alpha(u, h(v))$ takes a value between u and $h(v)$.

If the strict convexity conditions in (E_3) are satisfied along the local equilibria curve, then by continuity they will also be satisfied in the open domain D_l (possibly smaller).

Differentiating the Cauchy data (15.1.7) with respect to v leads to the identities

$$l'(v) = \eta_v(v,h(v)) + \eta_u(v,h(v))h'(v) = \eta_v(v,h(v)). \qquad (15.1.8)$$

Notice that in the second part of (15.1.7), there hold

$$l''(v) = \eta_{vv}(v, h(v)) + \eta_{vu}(v, h(v))h'(v), \qquad (15.1.9)$$

and

$$\eta_{uv}(v, h(v)) + \eta_{uu}(v, h(v))h'(v) = 0. \qquad (15.1.10)$$

From (15.1.10) and (15.1.9), we have

$$\eta_{uv}(v, h(v)) = \quad \eta_{uu}(v, h(v))h'(v)$$

and

$$\begin{aligned}
\eta_{vv}(v, h(v)) &= l''(v) - \eta_{vu}(v, h(v))h'(v) \\
&= l''(v) + \eta_{uu}(v, h(v))(h'(v))^2,
\end{aligned}$$

which combining with the entropy equation (E_1) yields that on $u = h(v)$, there holds

$$\begin{aligned}
0 &= \frac{1}{Z}\big(g_v\eta_{uu} - f_u\eta_{vv} + (f_v - g_u)\eta_{vu}\big) \\
&= \frac{1}{Z}\big(g_v - (f_v - g_u)h'(v) - f_u(h'(v))^2\big)\eta_{uu} - \frac{f_u}{Z}l''(v),
\end{aligned} \qquad (15.1.11)$$

or equivalently

$$\begin{aligned}
\frac{f_u^2}{Z^2}l''(v) &= \frac{1}{Z^2}\Big(f_u\big(g_v\eta_{uu} - f_u\eta_{vv} + (f_v - g_u)\eta_{vu}\big)\eta_{uu}\Big) \\
&= \frac{(\lambda_2(v, h(v)) - \lambda(v))(\lambda(v) - \lambda_1(v, h(v)))}{Z(v, u)^2}\eta_{uu}(v, h(v))
\end{aligned} \qquad (15.1.12)$$

and hence, $\eta_{uu}(v, h(v)) > 0$ from the conditions in (15.1.4).

To obtain the second part in (E_3), from (15.1.10), and

$$(f_v - g_u)\eta_{vu} = f_u\eta_{vv} - g_v\eta_{uu},$$

there holds

$$h'(v) = -\frac{\eta_{vu}(v, h(v))}{\eta_{uu}(v, h(v))}.$$

Substituting it into the identity (15.0.12), from the entropy equation (15.0.15), we obtain on $u = h(v)$ that

$$(\lambda_2(v, h(v)) - \lambda(v))(\lambda(v) - \lambda_1(v, h(v)))$$
$$= f_u(g_v + (g_u - f_v)h'(v) - f_u(h'(v))^2)$$
$$= f_u\left(g_v - (g_u - f_v)\frac{\eta_{vu}}{\eta_{uu}} - f_u\left(\frac{\eta_{vu}}{\eta_{uu}}\right)^2\right) \qquad (15.1.13)$$
$$= \frac{f_u^2}{\eta_{uu}^2}\left(\eta_{uu}\eta_{vv} - \left(\eta_{uv}\right)^2\right).$$

Therefore

$$\eta_{uu}\eta_{vv} - \left(\eta_{uv}\right)^2 > 0 \text{ on } u = h(v), \qquad (15.1.14)$$

which completes the proof of Lemma 15.1.2. ∎

Now we are in the position to give the main result in this section. Suppose that $(v^{\tau,\varepsilon}, u^{\tau,\varepsilon}) \in D_l$ are solutions of the Cauchy problem

$$\begin{cases} v_t + f(v, u)_x = \varepsilon v_{xx}, \\ u_t + g(v, u)_x + \frac{1}{\tau}(u - h(v)) = \varepsilon u_{xx}, \end{cases} \qquad (15.1.15)$$

with bounded initial data

$$(v, u)|_{t=0} = (v_0(x), u_0(x)) \in D_l, \qquad (15.1.16)$$

where D_l is given in Lemma 15.1.2 and

$$(v^{\tau,\varepsilon}, u^{\tau,\varepsilon}) \to (v^\tau, u^\tau) \quad a.e. \text{ as } \varepsilon \to 0, \qquad (15.1.17)$$

and the limit (v^τ, u^τ) is a weak solution of the Cauchy problem (15.0.1)-(15.0.2) in the domain D_l. Then we have the main result in this section given by the following theorem:

Theorem 15.1.3 *If the conditions in Lemma 15.1.2 are satisfied and*

$$meas\{v : \lambda(v) = 0\} = 0,$$

then there exists a subsequence (still denoted) (v^τ, u^τ) such that

$$(v^\tau, u^\tau) \to (v, u) \quad a.e.,$$

and the limit functions (v, u) satisfy
(i) $u(x, t) = h(v(x, t))$ a.e., for $t > 0$;
(ii) $v(x, t)$ is the weak solution of the Cauchy problem (15.0.6) with the initial data $v(x, 0) = v_0(x)$.

Proof. Let $\eta(v,u)$ be the strictly convex entropy of system (15.0.1) constructed in Lemma 15.1.2, and $q(v,u)$ be the corresponding entropy flux. Then

$$\eta_{vv}a^2 + 2\eta_{vu}ab + \eta_{uu}b^2 \geq c_0(a^2 + b^2) \tag{15.1.18}$$

for any vector $(a,b) \neq (0,0)$ and a positive constant c_0.

Since $\eta_{uu}(v,u) \geq d_1 > 0$ and $\eta_u(v, h(v)) = 0$, then in D_l, we have

$$\eta_u(v,u)(u - h(v)) = \eta_{uu}(v, \alpha(u, h(v)))(u - h(v))^2 \geq d_1(u - h(v))^2, \tag{15.1.19}$$

where $\alpha(u, h(v))$ denotes a value between u and $h(v)$.

Multiplying system (15.0.1) by (η_v, η_u), we obtain

$$\eta_t + q_x + \frac{\eta_u}{\tau}(u - h(v))$$

$$= \varepsilon\eta_{xx} - \varepsilon\left(\eta_{vv}v_x^2 + 2\eta_{vu}v_xu_x + \eta_u u_x^2\right). \tag{15.1.20}$$

Use (15.1.18) and (15.1.19) to obtain

$$\eta_t + q_x + \frac{d_1}{\tau}(u - h(v))^2 + \varepsilon c_0\left(v_x^2 + u_x^2\right) \leq \varepsilon\eta_{xx}. \tag{15.1.21}$$

Multiplying (15.1.21) by a suitable test function and then integrating in $R \times R^+$, we have that

$$\varepsilon\left(\left(v_x^{\tau,\varepsilon}\right)^2 + \left(u_x^{\tau,\varepsilon}\right)^2\right) \quad \in L^1_{loc}(R \times R^+) \tag{15.1.22}$$

and

$$\frac{1}{\tau}\left(u^{\tau,\varepsilon} - h\left(v^{\tau,\varepsilon}\right)\right)^2 \quad \in L^1_{loc}(R \times R^+). \tag{15.1.23}$$

Let ε tend to zero in (15.1.23) to obtain

$$\frac{1}{\tau}\left(u^{\tau} - h\left(v^{\tau}\right)\right)^2 \quad \in L^1_{loc}(R \times R^+). \tag{15.1.24}$$

Let $l_1(v) = v, l_2(v) = f(v, h(v)) = f(v)$ and related corresponding fluxes of the equilibrium equation (15.0.6) be $L_1(v), L_2(v)$. Let the entropies of system (15.0.1) with initial data $l(v) = l_1(v), l(v) = l_2(v)$ be $\eta_1(v,u), \eta_2(v,u)$ with corresponding entropy fluxes $q_1(v,u), q_2(v,u)$, respectively.

Multiplying system (15.0.1) by $(\eta_{iv}, \eta_{iu}), i = 1, 2$, we obtain

$$
\eta_i\left(v^{\tau,\varepsilon}, u^{\tau,\varepsilon}\right)_t + q_i\left(v^{\tau,\varepsilon}, u^{\tau,\varepsilon}\right)_x + \frac{\eta_{iu}}{\tau}\left(u^{\tau,\varepsilon} - h\left(v^{\tau,\varepsilon}\right)\right)
$$
$$
= \varepsilon\eta_{ixx} - \varepsilon\left(\eta_{ivv}\left(v_x^{\tau,\varepsilon}\right)^2 + 2\eta_{iuv}v_x^{\tau,\varepsilon}u_x^{\tau,\varepsilon} + \eta_{iuu}\left(u_x^{\tau,\varepsilon}\right)^2\right).
$$
$$
(15.1.25)
$$

Then

$$
l_i(v^{\tau,\varepsilon})_t + L_i(v^{\tau,\varepsilon})_x
$$
$$
= \left(l_i(v^{\tau,\varepsilon} - \eta_i\left(v^{\tau,\varepsilon}, u^{\tau,\varepsilon}\right)\right)_t + \left(L_i(v^{\tau,\varepsilon} - q\left(v^{\tau,\varepsilon}, u^{\tau,\varepsilon}\right)\right)_x
$$
$$
- \frac{\eta_{iuu}\left(v^{\tau,\varepsilon}, \alpha\right)}{\tau}\left(u^{\tau,\varepsilon} - h\left(v^{\tau,\varepsilon}\right)\right)^2 + \varepsilon\eta_{ixx}
$$
$$
- \varepsilon\left(\eta_{ivv}\left(v_x^{\tau,\varepsilon}\right)^2 + 2\eta_{iuv}v_x^{\tau,\varepsilon}u_x^{\tau,\varepsilon} + \eta_{iuu}\left(u_x^{\tau,\varepsilon}\right)^2\right),
$$
$$
(15.1.26)
$$

where α takes a value between $u^{\tau,\varepsilon}$ and $h\left(v^{\tau,\varepsilon}\right)$.

From (15.1.22), $\varepsilon\eta_{ixx}\left(v^{\tau,\varepsilon}, u^{\tau,\varepsilon}\right)$ tends to zero in the sense of distributions as ε tends to zero. Then letting $\varepsilon \to 0$ in (15.1.26), we have

$$
l_i(v^{\tau})_t + L_i(v^{\tau})_x = I_{i1}^{\tau} + I_{i2}^{\tau} + I_{i3}^{\tau} + I_{i4}^{\tau} \qquad (15.1.27)
$$

in the sense of distributions, where $I_{i3}^{\tau}, I_{i4}^{\tau}$ are weak limits of

$$
- \frac{\eta_{iuu}\left(v^{\tau,\varepsilon}, \alpha\right)}{\tau}\left(u^{\tau,\varepsilon} - h\left(v^{\tau,\varepsilon}\right)\right)^2
$$

and

$$
- \varepsilon\left(\eta_{ivv}\left(v_x^{\tau,\varepsilon}\right)^2 + 2\eta_{iuv}v_x^{\tau,\varepsilon}u_x^{\tau,\varepsilon} + \eta_{iuu}\left(u_x^{\tau,\varepsilon}\right)^2\right)
$$

as $\varepsilon \to 0$, being bounded in $L^1_{loc}(R \times R^+)$ from (15.1.22)-(15.1.23) and hence, compact in $W^{-1,p}_{loc}(R \times R^+)$ for $1 < p < 2$. Moreover

$$
\|I_{i1}^{\tau}\| = \sup_{\phi\in H_0^1} |\int\int \left(l_i(v^{\tau}) - \eta_i(v^{\tau}, u^{\tau})\right)_t \phi dx dt|
$$
$$
\leq C\|u^{\tau} - h(v^{\tau})\|_{L^2}\|\phi_t\|_{L^2}
$$
$$
\leq \sqrt{\tau}C\|\phi_t\|_{H^1} \to 0, \text{ as } \tau \to 0
$$
$$
(15.1.28)
$$

and

$$\|I_{i2}^\tau\| = \sup_{\phi \in H_0^1} |\int \int \left(L_i(v^\tau) - q_i(v^\tau, u^\tau) \right)_t \phi \, dx \, dt|$$

$$= \sup_{\phi \in H_0^1} |\int \int \left(q_i(v^\tau, h(v^\tau)) - q_i(v^\tau, u^\tau) \right)_t \phi \, dx \, dt| \quad (15.1.29)$$

$$\leq C \|u^\tau - h(v^\tau)\|_{L^2} \|\phi_x\|_{L^2}$$

$$\leq \sqrt{\tau} C \|\phi_t\|_{H^1} \to 0, \text{ as } \tau \to 0.$$

These imply that I_{i1}^τ, I_{i2}^τ are compact in $W_{loc}^{-1,2}(R \times R^+)$, and hence

$$l_i(v^\tau)_t + L_i(v^\tau)_x$$

is compact in $W_{loc}^{-1,p}(R \times R^+)$ for $1 < p < 2$. From the boundedness of (v^τ, u^τ), we conclude that

$$l_i(v^\tau)_t + L_i(v^\tau)_x$$

is bounded in $W_{loc}^{-1,\infty}(R \times R^+)$, and hence

$$l_i(v^\tau)_t + L_i(v^\tau)_x$$

is compact in $W_{loc}^{-1,2}(R \times R^+)$ from Theorem 2.3.2.

Therefore the compactness framework given in Chapter 3 shows the strong convergence $v^\tau \to v, a.e.$, which implies also the strong convergence $u^\tau \to u, a.e.$ from (15.1.24). So we get the proof of Theorem 15.1.3. ■

15.2 System of Extended Traffic Flows

In this section, we shall introduce an application of Theorem 15.1.3 on the relaxation problem for the nonstrictly hyperbolic system of extended traffic flows (15.0.17) with the initial data

$$(\rho(x,0), u(x,0)) = (\rho_0(x), u_0(x)) \quad (\rho_0(x) \geq 0). \quad (15.2.1)$$

Two eigenvalues of system (15.0.17) are

$$\lambda_1 = u - \sqrt{\rho p'(\rho)} \quad \lambda_2 = u + \sqrt{\rho p'(\rho)},$$

and the first-order relaxation correction corresponding to (15.0.11) is

$$\rho_t + (\rho h(\rho))_x = \tau\big(\phi(\rho)\rho_x\big)_x, \tag{15.2.2}$$

where $\phi(\rho) = \rho^2\big(\frac{p'(\rho)}{\rho} - (h'(\rho))^2\big)$, and hence the condition in (15.0.14) is reduced to

$$\frac{p'(\rho)}{\rho} - (h'(\rho))^2 > 0. \tag{15.2.3}$$

The entropy equation of system (15.0.17) is

$$\frac{p'(\rho)}{\rho}\eta_{uu} - \eta_{\rho\rho} = 0 \tag{15.2.4}$$

since $f_u(\rho, u) = \rho$, $g_\rho(\rho, u) = p'(\rho)$ and $f_\rho(\rho, u) - g_u(\rho, u) = 0$ have a common zero factor ρ if we assume $\frac{p'(\rho)}{\rho} \geq d > 0$. Therefore combining Theorem 15.1.3, Theorem 10.0.2 and Theorem 14.3.3 yields the following theorem:

Theorem 15.2.1 *Let $p_1(\rho) = \frac{p'(\rho)}{\rho} \geq d > (h'(\rho))^2$ for a positive constant d and $p'_1(\rho) \geq 0$. Suppose that there exist two small constants N, L such that the curve $u = h(\rho)$ passes the unique intersection point $(\bar{\rho}, \bar{u})$ of curves $w = N, z = -L$; the curve $u = h(\rho)$ and the initial data $(\rho_0(x), u_0(x))$ are in the region $\Sigma_9 = \{(\rho, u) : w \leq N, z \geq -L, \rho \geq 0\}$ as $0 \leq \rho \leq \bar{\rho}$ (see Figure 14.2). Then, for any fixed τ, the global weak solution (ρ^τ, u^τ) of the Cauchy problem (15.0.17), (15.2.1) exists. Moreover, if*

$$meas\ \{\rho : (\rho h(\rho))'' = 0\} = 0,$$

then there exists a subsequence (still denoted) (ρ^τ, u^τ) such that

$$(\rho^\tau, u^\tau) \to (\rho, u) \quad a.e.,$$

and the limit functions (ρ, u) satisfy

(i) $u(x, t) = h(\rho(x, t))$, a.e., for $t > 0$;

(ii) $\rho(x, t)$ is the unique weak solution of the Cauchy problem to the scalar equation

$$\rho_t + (\rho h(\rho))_x = 0$$

with the initial data

$$\rho(x, 0) = \rho_0(x).$$

15.3 Related Results

The general compactness framework to the singular limits of the general 2×2 strictly hyperbolic system (15.0.1) with the stiff relaxation term is established in [CLL], where an application on singular limits of relaxation approximated solutions to the system of elasticity is also obtained.

The extension, Theorem 15.1.3 of Chen-Levermore-Liu's framework to nonstrictly hyperbolic systems of two equations with a relaxation term, and its application on the extended model of traffic flow (15.0.17) both are established by Lu [Lu10].

Chapter 16

Relaxation for 3×3 Systems

In this chapter, we study the singular limit for the following nonlinear systems of three equations:

$$\begin{cases} v_t - u_x = 0, \\ u_t - \sigma(v, s)_x = 0, \\ s_t + c_1 s_x + \beta \dfrac{s - h(v)}{\tau} = 0, \end{cases} \qquad (16.0.1)$$

with initial data

$$\left. (v, u, s) \right|_{t=0} = (v_0, u_0, s_0), \qquad (16.0.2)$$

where β, τ and c_1 are nonnegative constants. When $\beta = 0$, the existence of L^2 global weak solution for the Cauchy problem (16.0.1), (16.0.2) is obtained in Section 12.3.

In the case of $\beta \neq 0$, for instance, $\beta = 1$, when written in Eulerian coordinates, system (16.0.1) can be used to model the chemically reacting flow (cf. [LLL]). Here v is specific volume, u denotes velocity, s is the mass fraction of one mode of the two-mode gas and $h(v)$ is the given equilibrium distribution in v. In this case, τ denotes the reaction time or relaxation time.

System (16.0.1) arises also in many physical situations, such as the system of adiabatic gas flow through porous media, the variant system of Broadwell model [LK2] and the system of isothermal motions of a viscoelastic material [LWu, Tz].

Formally, when the relaxation time $\tau \to 0$, it yields the system of elasticity or equations of isothermal elastodynamics,

$$\begin{cases} v_t - u_x = 0, \\ u_t - \sigma(v, h(v))_x = 0. \end{cases} \tag{16.0.3}$$

The three eigenvalues of system (16.0.1) are

$$\lambda_1 = -\sqrt{\sigma_v(v, s)}, \quad \lambda_2 = \sqrt{\sigma_v(v, s)}, \quad \lambda_3 = c_1 \tag{16.0.4}$$

and two eigenvalues of system (16.0.3) are

$$\lambda_1 = -\sqrt{\frac{d\sigma(v, h(v))}{dv}}, \quad \lambda_2 = \sqrt{\frac{d\sigma(v, h(v))}{dv}}. \tag{16.0.5}$$

In Section 16.1, we are interested in combining the zero relaxation with the zero dissipation limit of the Cauchy problem

$$\begin{cases} v_t - u_x = \varepsilon v_{xx}, \\ u_t - \sigma(v, s)_x = \varepsilon u_{xx}, \\ s_t + c_1 s_x + \beta \frac{s - h(v)}{\tau} = \varepsilon s_{xx}, \end{cases} \tag{16.0.6}$$

with the initial data (16.0.2).

When $\tau \le \frac{\epsilon}{M}$, with M a suitable large constant which depends only on the initial data, and ϵ goes to zero, the convergence of the solutions $\left(v^{\epsilon,\tau}, u^{\epsilon,\tau}, s^{\epsilon,\tau}\right)$ to the Cauchy problem (16.0.6), (16.0.2) is obtained for very general $\sigma(v, s)$ even if (16.0.1) is a elliptic-hyperbolic mixed system, that is, $\sigma_v(v, s) \not\ge 0$.

In Section 16.2, we consider the relaxation limit for the following special case of system (16.0.1) without introducing the viscosity:

$$\begin{cases} v_t - u_x = 0, \\ u_t - (v - cs)_x = 0, \\ s_t + \frac{s - h(v)}{\tau} = 0, \end{cases} \tag{16.0.7}$$

with the initial data (16.0.2), where c is a positive constant, but the nonlinear function $h(v)$ must satisfy the subcharacteristic condition

$$0 < d_1 \le h'(v) \le d_2 < \frac{1}{c} \tag{16.0.8}$$

for positive constants d_1 and d_2.

16.1   Dominant Diffusion and Stiff Relaxation

In this section, we shall study the singular limits of stiff relaxation and dominant diffusion for the Cauchy problem of (16.0.6), (16.0.2), that is, the relaxation time τ tends to zero faster than the diffusion parameter ϵ, $\tau = o(\epsilon), \epsilon \to 0$. We establish the following general compactness framework:

Theorem 16.1.1 <u>A:</u> *If the initial data* (v_0, u_0, s_0) *are smooth functions satisfying the following condition:*

($\mathbf{c_1}$) $\left| v_0, u_0, s_0 \right|_{L^2 \cap L^\infty (R)} \leq M_1$

$$\lim_{|x| \to \pm\infty} \left(\frac{d^i v_0}{dx^i}, \frac{d^i u_0}{dx^i}, \frac{d^i s_0}{dx^i} \right) = (0, 0, 0), \quad i = 0, 1;$$

($\mathbf{c_2}$) $h(v) = cv$, $\sigma(v, s)$ *satisfies the following condition:*

$$\left| \sigma_s(v, s) \right| \leq M_2, \quad \bar{\sigma}'(v) \geq d > \max\{0, c^2 - c + \frac{2c^2 c_1^2}{(M_2 + 1)^2}\},$$

where $\bar{\sigma}(v) = \sigma(v, cv)$, *then for fixed* ϵ, τ *satisfying* $\tau(M_2 + 1)^2 \leq \epsilon$, *the solutions* $(u, v, s) \in C^2$ *of the Cauchy problem (16.0.6), (16.0.2) exist in* $(-\infty, \infty) \times [0, T]$ *for any given* $T > 0$ *and satisfy*

$$|v(x,t)|, \quad |u(x,t)|, \quad |s(x,t)| \quad \leq M(\epsilon, \tau, T), \tag{16.1.1}$$

$$|v(\cdot,t)|_{L^2(R)}, \quad |u(\cdot,t)|_{L^2(R)}, \quad |s(\cdot,t)|_{L^2(R)} \quad \leq M, \tag{16.1.2}$$

$$|(s - cv)^2|_{L^1(R \times R^+)} < \tau M, \tag{16.1.3}$$

$$|\epsilon v_x^2|_{L^1(R \times R^+)}, \quad |\epsilon u_x^2|_{L^1(R \times R^+)}, \quad |\epsilon s_x^2|_{L^1(R \times R^+)} \leq M.$$

B: *If* $\bar{\sigma}(v) = \sigma(v, cv)$ *satisfies the following condition:*

($\mathbf{c_3}$) $\bar{\sigma}''(v_0) = 0$ *and* $\bar{\sigma}''(v) \neq 0$ *for* $v \neq v_0$, $\bar{\sigma}''$, $\bar{\sigma}''' \in L^2 \cap L^\infty$, *then there exist a subsequence (still denoted by)* $(v^{\epsilon,\tau}, u^{\epsilon,\tau}, s^{\epsilon,\tau})$ *of the solutions to the Cauchy problem (16.0.6), (16.0.2) and* L^2 *bounded functions* (v, u, s) *such that*

$$\left(v^{\epsilon,\tau}, u^{\epsilon,\tau}, s^{\epsilon,\tau} \right) \to (v, u, s) \quad a.e.(x,t), \tag{16.1.4}$$

where (v, u, s) *satisfies* $s = h(v)$ *and* (v, u) *is an entropy solution of the equilibrium system (16.0.3) with the initial data* $(v_0(x), u_0(x))$.

Remark 16.1.2 *In Theorem 16.1.1, the condition $h(v) = cv$ is for avoiding the technical details. We shall see from the proof of Theorem 16.1.1 below that all the steps work just as well for a more general function $h(v)$ which satisfies $h'(v) \geq d_1 > 0$. In fact, we only need to write the term $s - h(v)$ as a different form:*

$$s - h(v) = h'(\theta)(h^{-1}(s) - v) \tag{16.1.5}$$

where h^{-1} is the inverse function of h and θ takes a value between $h^{-1}(s)$ and v.

Proof of Theorem 16.1.1. To prove part A, we use the following local existence lemma and the L^∞ estimates given in (16.1.1).

Lemma 16.1.3 *(Local existence) If the initial data satisfies condition $(\mathbf{c_1})$ in Theorem 16.1.1, then for any fixed ϵ and $\tau > 0$, the Cauchy problem (16.0.6), (16.0.2) admits a unique smooth local solution (u, v, s) which satisfies*

$$|\frac{\partial^i v}{\partial x^i}| + |\frac{\partial^i u}{\partial x^i}| + |\frac{\partial^i s}{\partial x^i}| \leq M(t_1, \epsilon, \tau) < +\infty \quad i = 0, 1, 2 \tag{16.1.6}$$

where $M(t_1, \epsilon, \tau)$ is a positive constant that depends only on t_1, ϵ, τ and t_1 depends on $|v_0|_{L^\infty}, |u_0|_{L^\infty}, |s_0|_{L^\infty}$.

Moreover

$$\lim_{|x| \to \pm\infty} \left(\frac{d^i v}{dx^i}, \frac{d^i u}{dx^i}, \frac{d^i s}{dx^i} \right) = (0, 0, 0), \quad i = 0, 1 \tag{16.1.7}$$

uniformly in $t \in [0, t_1]$.

Proof. Lemma 16.1.2 can be proved by applying the Banach contraction mapping theorem to an integral representation of (16.0.6). For details see Theorem 1.0.2. ∎

To derive the crucial estimates given in (16.1.1), we need the necessary condition $\tau(M_2 + 1)^2 \leq \epsilon$ and condition (c_2) in Theorem 16.1.1.

Multiplying the first equation in (16.0.6) by $\bar{\sigma}(v) + cv - cs$, the second by u and the third by $s - cv$ and adding the result, we have

$$
\left(\int_0^v \bar{\sigma}(v) + cv dv + \frac{u^2}{2} - csv + \frac{s^2}{2} \right)_t
$$
$$
+ \left(cus - u(\bar{\sigma}(v) + cv) + \frac{c_1 s^2}{2} \right)_x
$$
$$
- cc_1 vs_x - u \Big(\sigma(v,s) + s - (\sigma(v,cv) + cv) \Big)_x + \frac{(s-cv)^2}{\tau}
$$
$$
= \varepsilon \left(\int_0^v \bar{\sigma}(v) + cv dv + \frac{u^2}{2} - csv + \frac{s^2}{2} \right)_{xx}
$$
$$
- \varepsilon \big(\bar{\sigma}'(v) + c \big) v_x^2 - \varepsilon u_x^2 - \varepsilon s_x^2 + 2c \varepsilon s_x v_x.
$$

$$(16.1.8)$$

For the third and fourth terms on the left-hand side of (16.1.8), we have the estimate

$$
- cc_1 vs_x - u\Big(\sigma(v,s) + s - (\sigma(v,cv) + cv)\Big)_x
$$
$$
= \left(\frac{c_1 c^2}{2} v^2 - cc_1 vs \right)_x - \Big(u(\sigma(v,s) + s - (\sigma(v,cv) + cv)) \Big)_x
$$
$$
+ u_x \big(\sigma_s(v,\alpha) + 1 \big)(s - cv) + cc_1 v_x (s - cv)
$$

$$(16.1.9)$$

where α takes a value between s and cv. The last two terms in (16.1.9) have the upper bound

$$
\frac{3(s-cv)^2}{4\tau} + \frac{\tau(M_2+1)^2 u_x^2}{2} + \tau c^2 c_1^2 v_x^2 \qquad (16.1.10)
$$

by the first condition in ($\mathbf{c_2}$).

Combining (16.1.8), (16.1.9) and (16.1.10), we get the following inequality:

$$
\left(\int_0^v \bar{\sigma}(v) + cv dv + \frac{u^2}{2} - csv + \frac{s^2}{2} \right)_t
$$
$$
+ \left(cus - u(\bar{\sigma}(v) + cv) + \frac{c_1}{2} s^2 \right)_x - \left(cc_1 vs - \frac{c_1 c^2}{2} v^2 \right)_x
$$
$$
- \Big(u(\sigma(v,s) + s - (\sigma(v,cv) + c)) \Big)_x + \frac{(s-cv)^2}{4\tau}
$$
$$
\leq \varepsilon \left(\int_0^v (\bar{\sigma}(v) + cv) dv + \frac{u^2}{2} - csv + \frac{s^2}{2} \right)_{xx} - \varepsilon \big(\bar{\sigma}'(v) + c \big) v_x^2
$$
$$
- \varepsilon s_x^2 + 2c \varepsilon s_x v_x + c^2 c_1^2 \tau v_x^2 - \left(\varepsilon - \frac{\tau(M_2+1)^2}{2} \right) u_x^2.
$$

$$(16.1.11)$$

Noticing the second condition in (c_2) we know that

$$\int_0^v (\bar{\sigma}(v) + cv)dv + \frac{u^2}{2} - csv + \frac{s^2}{2}$$

is a strictly convex function. If the condition $\tau(M_2 + 1)^2 \le \epsilon$ is satisfied, noticing (16.1.7), we immediately get the estimates (16.1.2), (16.1.3) by integrating (16.1.11) on $R \times [0, T]$.

Differentiating the first equation in (16.0.6) with respect to x, we get

$$\left(v_x\right)_t - u_{xx} = \varepsilon\left(v_x\right)_{xx}. \tag{16.1.12}$$

Multiplying (16.1.12)by v_x yields

$$\left(\frac{v_x^2}{2}\right)_t - \left(v_x u_x\right)_x + u_x v_{xx} = \varepsilon\left(\frac{v_x^2}{2}\right)_{xx} - \varepsilon v_{xx}^2. \tag{16.1.13}$$

Integrating (16.1.13) in $R \times [0, T]$ and noticing $|u_x^2|_{L^1(R \times R^+)} \le M(\epsilon)$, we obtain the bound $|v_x^2(\cdot, t)|_{L^1(R)} \le M(\epsilon)$, where $M(\epsilon)$ is a constant depending on ϵ. Therefore

$$v^2 = |\int_{-\infty}^x (v^2)_x dx| \le \int_{-\infty}^\infty v^2 dx + \int_{-\infty}^\infty v_x^2 dx \le M(\epsilon).$$

Similarly, we can get $|s_x^2(\cdot, t)|_{L^1(R)} \le M(\epsilon)$ and $|u_x^2(\cdot, t)|_{L^1(R)} \le M(\epsilon)$ from the second and third equations in (16.0.6). So we get the estimates in (16.1.1) and hence the proof of (A) in Theorem 16.1.1.

From the estimates in (16.1.2) and (16.1.3), it is easy to prove the compactness of $\eta(v^{\epsilon,\tau})_t + q(v^{\epsilon,\tau})_x$ in $H_{loc}^{-1}(R \times R^+)$, where $(v^{\epsilon,\tau}, u^{\epsilon,\tau})$ are the solutions of the Cauchy problem (16.0.6), (16.0.2) and (η, q) is any entropy-entropy flux pair of Shearer type constructed in Section 12.2. Then the convergence of $(v^{\epsilon,\tau}, u^{\epsilon,\tau})$ follows. From the first estimate in (16.1.3), we obtain the convergence $s^{\epsilon,\tau} \to s$. So Theorem 16.1.1 is proved. ∎

16.2   A Model System for Reacting Flow

In this section we shall consider the stiff relaxation limit of solutions for the system of chemically reacting flow (16.0.7) with the initial data (16.0.2).

We establish the following theorem:

Theorem 16.2.1 *Let the condition (16.0.8) hold. Then, for any fixed τ, the Cauchy problem (16.0.7), (16.0.2) has a unique global smooth solution (v^τ, u^τ, s^τ). Furthermore, suppose the initial data satisfy*

$$\int_R v_0^2 + u_0^2 + s_0^2 dx \le M, \quad \tau^2 \int_R v_{0x}^2 + u_{0x}^2 + s_{0x}^2 dx \le M, \quad (16.2.1)$$

and

$$h''(v_0) = 0, \quad h''(v) \ne 0 \text{ for } v \ne v_0, \quad (16.2.2)$$

$$h''(v) \in L^2(R) \cap L^\infty(R), \quad h'''(v) \in L^2(R) \cap L^\infty(R). \quad (16.2.3)$$

Then, along a subsequence if necessary,

$$\left(v^\tau, u^\tau, s^\tau\right) \to (v, u, s) \quad a.c. \ (x, t), \quad (16.2.4)$$

where (v, u, s) satisfies $s = h(v)$ and (v, u) is an entropy solution of the equilibrium system

$$\begin{cases} v_t - u_x = 0, \\ u_t - (v - ch(v))_x = 0, \end{cases} \quad (16.2.5)$$

with the initial data $\left(v_0(x), u_0(x)\right)$.

The right side of condition (16.0.8) ensures the equilibrium system (16.2.5) is strictly hyperbolic since in this case, $\frac{d(v-ch(v))}{dv} = 1 - ch'(v) > 0$; and the left side of (16.0.8) is equivalent to the strictly subcharacteristic condition

$$\lambda_1 < \bar\lambda_1 < \lambda_2 < \bar\lambda_2 < \lambda_3, \quad (16.2.6)$$

where $\lambda_1 = -1, \lambda_2 = 0, \lambda_3 = 1$ are three eigenvalues of system (16.0.7), and

$$\bar\lambda_1 = -\sqrt{1 - ch'(v)}, \quad \bar\lambda_2 = \sqrt{1 - ch'(v)}$$

are two eigenvalues of equilibrium system (16.2.5).

We now show through an asymptotic expansion of the Chapman-Enskog type that condition (16.0.8) is necessary to ensure the stability of the relaxation approximated solutions (v^τ, u^τ, s^τ) of the Cauchy problem (16.0.7), (16.0.2).

Let $s^\tau = h(v^\tau) + \tau g(v^\tau, u^\tau, v_x^\tau, u_x^\tau, \ldots) + O(\tau^2)$.

To calculate the form of g, we use system (16.0.7),

$$
\begin{cases}
v_t^\tau - u_x^\tau = 0, \\
u_t^\tau - (v^\tau - ch(v^\tau))_x = -\tau g_x + O(\tau^2), \\
h(v^\tau)_t + O(\tau) = -g + O(\tau).
\end{cases}
\tag{16.2.7}
$$

Substituting g from the third equation of (16.2.7) into the second and noticing that $v_t^\tau = u_x^\tau$ from the first equation, we have

$$
\begin{cases}
v_t^\tau - u_x^\tau = 0, \\
u_t^\tau - (v^\tau - ch(v^\tau))_x = \tau(h'(v^\tau)u_x^\tau)_x + O(\tau^2),
\end{cases}
\tag{16.2.8}
$$

which is a stable hyperbolic-parabolic mixed system, or the so-called Navier-Stokes equations if $h'(v) \geq d_1 > 0$.

For fixed $\tau > 0$, the existence of the unique global smooth solution (v^τ, u^τ, s^τ) in Theorem 16.2.1 is obvious since the growth order of the unique nonlinear function $h(v)$ is given by (16.0.8).

We shall prove the compactness (16.2.4) in Theorem 16.2.1 in several steps by several lemmas.

Lemma 16.2.2 *Under condition (16.0.8), there holds*

$$
\int_R (v^2 + u^2 + s^2)dx + \frac{1}{\tau C} \int_0^t \int_R (s - h(v))^2 dx dt
$$

$$
\leq C \int_R (v_0^2 + u_0^2 + s_0^2)dx,
\tag{16.2.9}
$$

for some C independent of τ and t.

Proof. Let the inverse function of h be h^{-1}. Then $(h^{-1})' = \frac{1}{h'} \in [\frac{1}{d_2}, \frac{1}{d_1}]$ and hence $h^{-1}(s) - v = (h^{-1})'(\alpha)(s - h(v))$ for a value α between s and $h(v)$.

Multiplying the first equation in (16.0.7) by $(v - cs)$, the second by u and the third by $c(h^{-1}(s) - v)$, we get

$$
(v^2 + u^2 - 2cvs + 2c \int_0^s h^{-1}(s)ds)_t - 2((v - cs)u)_x
$$

$$
+ \frac{2c(h^{-1})'(\alpha)}{\tau}(s - h(v))^2 = 0.
\tag{16.2.10}
$$

Since $2c \int_0^s h^{-1}(s)ds \geq cs^2 \min\{\frac{1}{h'}\} \geq \frac{cs^2}{d_2} > c^2 s^2$, then integrating (16.2.10) in $R \times [0, t]$, we get the proof of Lemma 16.2.2. ■

Using the second and third equations in (16.0.7), we have

$$
\begin{aligned}
u_t - (v - ch(v))_x &= -c(s - h(v))_x \\
&= cs_{xt} = (v_x - u_t)_t = \tau(u_{xx} - u_{tt}),
\end{aligned}
$$

(16.2.11)

and hence the following system:

$$
\begin{cases}
v_t - u_x = 0, \\
u_t - g(v)_x = \tau(u_{xx} - u_{tt}),
\end{cases}
$$

(16.2.12)

where $g(v) = v - ch(v)$.

Lemma 16.2.3 *Suppose the iniliul data satisfy (16.2.1). Then solutions (v, u, s) of (16.0.7) satisfy the τ independent estimates*

$$
\tau \int_0^t \int_R (v_x^2 + u_x^2 + s_x^2) dx dt \leq M.
$$

(16.2.13)

Proof. Multiplying the first equation in (16.2.12) by $g(v)$, the second by u, we have

$$
\left(\int_0^v g(v)dv + \frac{u^2}{2} \right)_t - (ug(v))_x = \tau \left(\frac{u^2}{2} \right)_{xx} - \tau u_x^2 - \tau \left(\frac{u^2}{2} \right)_{tt} + \tau u_t^2,
$$

(16.2.14)

or equivalently

$$
\left(\int_0^v g(v)dv + \frac{u^2}{2} + \tau u u_t \right)_t - (ug(v))_x + \tau(u_x^2 - u_t^2) = \tau(uu_x)_x.
$$

(16.2.15)

The problem is that the term $u_x^2 - u_t^2$ is not positive definite. To compensate for that, we first multiply the second equation in (16.2.12) by u_t to obtain

$$
u_t^2 - g'(v)v_x v_t = \tau \left((u_t u_x)_x - \left(\frac{u_x^2 + u_t^2}{2} \right)_t \right)
$$

(16.2.16)

and, in turn

$$
\tau^2 (u_x^2 + u_t^2)_t + 2\tau(u_t^2 - g'(v)v_x u_t) = 2\tau^2 (u_t u_x)_x.
$$

(16.2.17)

Using the second equation in (16.2.12) again, we have

$$
\begin{aligned}
g'(v)v_x^2 &= v_x(u + \tau u_t)_t - \tau v_x u_{xx} \\
&= (v(u + \tau u_t)_t)_x - v(u + \tau u_t)_{tx} - \tau v_x u_{xx} \\
&= (v(u + \tau u_t))_{tx} - (v_t(u + \tau u_t))_x - \tau v_x u_{xx} \\
&\quad + v_t(u + \tau u_t)_x - (v(u + \tau u_t)_x)_t \\
&= -(v_t(u + \tau u_t))_x + (v_x(u + \tau u_t))_t \\
&\quad + u_x(u + \tau u_t)_x - \tau\left(\tfrac{v_x^2}{2}\right)_t,
\end{aligned}
\tag{16.2.18}
$$

where $u_x = v_t$ is used in the last term $\tau v_x u_{xx}$.

(16.2.18) yields

$$
\begin{aligned}
\tau^2\left(\frac{v_x^2}{2} - \frac{u_x^2}{2}\right)_t &- \tau(v_x(u + \tau u_t))_t + \tau(g'(v)v_x^2 - u_x^2) \\
&= -\tau(v_t(u + \tau u_t))_x.
\end{aligned}
\tag{16.2.19}
$$

Adding (16.2.15),(16.2.17) and (16.2.19), we obtain that

$$
\begin{aligned}
&\left(\tfrac{1}{2}(u + \tau u_t - \tau v_x)^2 + \tfrac{1}{2}\tau^2(u_t^2 + u_x^2) + \int_0^v g(v)dv\right)_t \\
&- (ug(v))_x + \tau(u_t^2 - 2g'(v)v_x u_t + g'(v)v_x^2) \\
&= \tau^2(u_t u_x)_x.
\end{aligned}
\tag{16.2.20}
$$

Using condition (16.0.8) again, we have $g'(v) = 1 - ch'(v) \in [1 - cd_2, 1 - cd_1]$ and hence

$$
u_t^2 - 2g'(v)v_x u_t + g'(v)v_x^2 \geq c_1(u_t^2 + v_x^2)
\tag{16.2.21}
$$

for a positive constant c_1.

Therefore, integrating (16.2.19) in $R \times [0, t]$ and noticing (16.2.21), we have

$$
\tau\int_0^t\int_R u_t^2 + v_x^2 \, dxdt \leq M, \quad \tau^2\int_R u_t^2 + u_x^2 \, dx \leq M.
\tag{16.2.22}
$$

Integrating (16.2.15) in $R \times [0, t]$ and using the estimates (16.2.9) and (16.2.22), we have

$$
\tau\int_0^t\int_R u_x^2 \, dxdt \leq M.
\tag{16.2.23}
$$

Rewriting the third equation in (16.0.7), then differentiating it with respect to x, we get

$$\tau(s_t)_x s_x + (s_x - h'(v)v_x)s_x = 0. \tag{16.2.24}$$

Integrating (16.2.24) in $R \times [0, t]$ and using the estimate about v_x^2 in (16.2.22), we get

$$\tau^2 \int_0^t s_x^2 dx \leq M, \quad \tau \int_0^t \int_R s_x^2 dx dt \leq M, \tag{16.2.25}$$

and hence, the proof of Lemma 16.2.3. ∎

Proof of Theorem 16.2.1. Let $(\eta(v, u), q(v, u))$ be any entropy-entropy flux pair of Shearer type constructed in Chapter 12 for the equilibrium system (16.2.5).

Multiplying system (16.2.12) by (η_v, η_u) and noticing the first equality in (16.2.11), we have

$$\begin{aligned}
\eta(v, u)_t + q(v, u)_x &= c\eta_v(h(v) - s) \\
&= c\big(\eta_v(h(v) - s)\big)_x - c(\eta_{vv}v_x + \eta_{vu}u_x)(h(v) - s) \\
&= I_1 + I_2.
\end{aligned} \tag{16.2.26}$$

Clearly from the estimates in Lemma 16.2.2,

$$I_1 = c\big(\eta_v(h(v) - s)\big)_x = c\Big(\tau^{\frac{1}{2}}\eta_v \frac{(h(v) - s)}{\tau^{\frac{1}{2}}}\Big)_x$$

is compact in $W_{loc}^{-1,2}$ and

$$I_2 = -c(\eta_{vv}v_x + \eta_{vu}u_x)(h(v) - s) = -c\tau^{\frac{1}{2}}(\eta_{vv}v_x + \eta_{vu}u_x)\frac{(h(v) - s)}{\tau^{\frac{1}{2}}}$$

is bounded in L_{loc}^1, and hence compact in $W_{loc}^{-1,\alpha}$ for a constant $\alpha \in (1, 2)$. Thus we have that $\eta(v^\tau, u^\tau)_t + q(v^\tau, u^\tau)_x$ is compact in $W_{loc}^{-1,2}(R \times R^+)$. Then Shearer's compactness framework in Chapter 12 shows the convergence

$$(v^\tau, u^\tau) \to (v, u), a.e.,$$

which implies the convergence $s^\tau \to s, a.e.$ by the estimate given in the second part of the left-hand side of (16.2.9). This completes the proof of Theorem 16.2.1. ∎

16.3   Related Results

The proof of Theorem 16.1.1 for general system (16.0.1) is from [Lu12]. Before it, two special systems in the form (16.0.1) are studied by Lu-Klingenberg in [LK1, LK2].

About the relaxation limits for system (16.0.7), an equivalent system in the following form:

$$\begin{cases} v_t - u_x = 0, \\ u_t - s_x = 0, \\ (s - cv)_t + \dfrac{s - h(v)}{\tau} = 0, \end{cases} \tag{16.3.1}$$

was first studied by I.-S. Liu and Y.-M. Wu [LWu] for smooth relaxation approximated solutions. Later, the general singular limits for L^p solutions were obtained by Tzavaras [Tz] by Shearer's basic framework coupled some ideas given in [LK1, LK2].

Bibliography

[Ad] R. A. Adams, *Sobolev Spaces*, Academic Press, New York-San Francisco-London, 1975.

[Ba1] J. M. Ball, *Convexity conditions and existence theorems in the nonlinear elasticity*, Arch. Rat. Mech. Anal., **63** (1977), 337-403.

[Ba2] J. M. Ball, *A version of the fundamental theorem for Young measures*, Proc. Conf. on Partial Differential Equations and Continuum Models of Phase Transitions, Nice 1988, D. Serre, ed., Springer-Verlag, Berlin - Heidelberg - New York, 1988.

[BB] S. Bianchini and A. Bressan, *Vanishing Viscosity Solutions of Nonlinear Hyperbolic Systems*, preprint, S.I.S.S.A., Trieste, 2001.

[Bi] A. V. Bitsadze, *Equations of Mixed Type*, Macmillan, New York, 1964.

[BC] A. Bressan and R. M. Colombo, *Unique solutions of* 2×2 *conservation laws with large data*, Indiana Univ. Math. J., **44** (1995), 677-725.

[Br] A. Bressan, *The unique limit of the Glimm scheme*, Arch. Rat. Mech. Anal., **130** (1995), 205-230.

[BS] A. Bressan and W. Shen, *Estimates for multicomponent chromatography with relaxation*, Discr. Cont. Dyn. Syst., **6** (2000), 21-38.

[BR] J. E. Broadwell, *Shock structure in a simple discrete velocity gas*, Phys. Fluids, **7** (1964), 1243–1247.

[Ca] R. E. Caflish, *Navier-Stokes and Boltzmann shock profiles for a model of gas dynamics*, Comm. Pure Appl. Math., **32** (1979), 521–554.

[CEMP] S. Caprino, R. Esposito, R. Marra and M. Pulvirenti, *Hydrodynamic limits of the Vlasov equation*, Comm. Partial. Diff. Eqs., **18** (1993), 805-820.

[Ce] C. Cercignani, *The Boltzmann Equations and Its Application*, Spring-Verlag, New York, 1988.

[Ch1] G.-Q. Chen, *Convergence of the Lax-Friedrichs scheme for isentropic gas dynamics*, Acta Math. Sci., **6** (1986), 75-120.

[Ch2] G.-Q. Chen, *The compensated compactness method and the system of isentropic gas dynamics*, Preprint MSRI-00527-91, Mathematical Sciences Research Institute, Berkeley, 1990.

[Ch3] G.-Q. Chen, *Hyperbolic system of conservation laws with a symmetry*, Commun. PDE, **16** (1991), 1461-1487.

[Ch4] G.-Q. Chen, *Relaxation limit for conservation laws*, Z. Angew. Math. Mech., **76** (1996), 381-384.

[CG] G.-Q. Chen and J. Glimm, *Global solutions to the compressible Euler equations with geometric structure*, Commun. Math. Phys., **180** (1996), 153-193.

[CK] G.-Q. Chen and P.-T. Kan, *Hyperbolic conservation laws with umbilic degeneracy I*, Arch. Rat. Mech. Anal., **130** (1995), 231-276.

[CL] G.-Q. Chen and P. LeFloch, *Compressible Euler equations with general pressure law and related equations*, Arch. Rat. Mech. Anal., **153** (2000), 221-259.

[CLL] G.-Q. Chen, C. D. Levermore and T.-P. Liu, *Hyperbolic conservation laws with stiff relaxation terms and entropy*, Comm. Pure Appl. Math., **47**(1994), 787-830.

[ChL] G.-Q. Chen and T.-P. Liu, *Zero relaxation and dissipation limits for hyperbolic conservation laws*, Comm. Pure Appl. Math., **46** (1993), 755-781.

[CL1] G.-Q. Chen and Y.-G. Lu, *A study on the applications of the theory of compensated compactness*, Chinese Science Bulletin, **33** (1988), 641–644.

[CL2] G.-Q. Chen and Y.-G. Lu, *Convergence of the approximation solutions to isentropic gas dynamics*, Acta Math. Sci., **10** (1990), 39-46.

[Chern] I.-L. Chern, *Long-time effect of relaxation for hyperbolic conservation laws*, Commun. Math. Phys., **172** (1995), 39-55.

[CCS] K. N. Chueh, C. C. Conley and J. A. Smoller, *Positive invariant regions for systems of nonlinear diffusion equations*, Indiana Univ. Math. J., **26** (1977), 372-411.

[Da] B. Dacorogna, *Weak continuity and weak lower semicontinuity of non-linear functionals*, Lecture Notes in Math, **922**, Springer-Verlag, Berlin-Heidelberg-New York, 1982.

[Da1] C. M. Dafermos, *Estimates for conservation laws with little viscosity*, SIAM J. Math. Anal., **18** (1987), 409-421.

[Da2] C. M. Dafermos, *Hyperbolic conservation laws in continuum physics*, Grundlehren der Mathematischen Wissenschaften, Vol. 325, Springer Verlag, Berlin-Heidelberg-New York, 2001.

[DCL1] X.-X. Ding, G.-Q. Chen and P.-Z. Luo, *Convergence of the Lax-Friedrichs schemes for the isentropic gas dynamics I-II*, Acta Math. Sci., **5** (1985), 415-432, 433-472.

[DCL2] X.-X. Ding, G.-Q. Chen and P.-Z. Luo, *Convergence of the fractional step Lax-Friedrichs scheme and Godunov scheme for the isentropic system of gas dynamics*, Commun. Math. Phys., **121** (1989), 63-84.

[Di1] R. J. DiPerna, *Global solutions to a class of nonlinear hyperbolic systems of equations*, Comm. Pure Appl. Math., **26** (1973), 1-28.

[Di2] R. J. DiPerna, *Convergence of the viscosity method for isentropic gas dynamics*, Commun. Math. Phys., **91** (1983), 1-30.

[Di3] R. J. DiPerna, *Convergence of approximate solutions to conservation laws*, Arch. Rat. Mech. Anal., **82** (1983), 27-70.

[Di4] R. J. DiPerna, *Measure-valued solutions to conservation laws*, Arch. Rat. Mech. Anal., **88** (1985), 223-270.

[Ea] S. Earnshaw, *On the mathematical theory of sound*, Philos. Trans., **150** (1858), 1150-1154.

[Ev] L. C. Evans, *Weak convergence methods for nonlinear partial differential equations*, CBMS 74, Am. Math. Soc., Providence, Rhode Island, 1990.

[ES] L. C. Evans and P. E. Souganidis, *A PDE approach to geometric optics for certain semilinear parabolic equations*, Indiana Univ. Math. J., **38** (1989), 141-172.

[Fi] P. Fife, *Dynamics of Internal Layers and Diffusive Interfaces*, CBMS-NSF Regional Conference Series in Applied Mathematics 53, SIAM, Philadelphia, 1988.

[FS1] H. Frid and M. Santos, *Nonstrictly hyperbolic systems of conservation laws of the conjugate type*, Commun. PDE, **19** (1994), 27-59.

[FS2] H. Frid and M. Santos, *The Cauchy problem for the system $\partial_t z + \partial_x(\bar{z}^\gamma) = 0$*, J. Diff. Eqs., **111** (1995), 340-359.

[Gl] J. Glimm, *Solutions in the large for nonlinear hyperbolic systems of equations*, Comm. Pure Appl. Math., **18** (1965), 95-105.

[He] A. Heibig, *Existence and uniqueneons for some hyperbolic systems of conservation laws*, Arch. Rat. Mech. Anal., **126** (1994), 79-101.

[Ho] E. Hoff, *The partial differential equation $u_t + uu_x = \varepsilon u_{xx}$* , Comm. Pure Appl. Math., **3** (1950), 201-230.

[IMPT] E. Isaacon, D. Marchesin, B. Plohr and B. Temple, *The Riemann problem near a hyperbolic singularity: the classification of solutions of quadratic Riemann problem (I)*, SIAM J. Appl. Math., **48** (1988), 1-24.

[IT] E. Isaacon and B. Temple, *The classification of solutions of quadratic Riemann problem (II)-(III)*, SIAM J. Appl. Math., **48** (1988), 1287-1301, 1302-1318.

[JPP] F. James, Y.-J. Peng and B. Perthame, *Kinetic formulation for chromatography and some other hyperbolic systems*, J. Math. Pure Appl., **74** (1995), 367-385.

[Ka] E. Kamke, *Differentialgleichungen, Lösungsmethoden und Lösungen: 1. Gewöhnliche Differentialgleichungen*, sixth ed., Akademische Verlagsanstalt, Leipzig (1959).

[Kan] P.T. Kan, *On the Cauchy problem of a* 2×2 *system of nonstrictly hyperbolic conservation laws*, Ph.D. thesis, New York Univ., 1989.

[KK] B. Keyfitz and H. Kranzer, *A system of nonstrictly hyperbolic conservation laws arising in elasticity*, Arch. Rat. Mech. Anal., **72** (1980), 219-241.

[KM] S. Klainerman and A. Majda, *Singular limits of quasilinear hyperbolic systems with large parameters and the incompressible limit of compressible fluids*, Comm. Pure Appl. Math., **34** (1981), 481-524.

[KL1] C. Klingenberg and Y.-G. Lu, *Cauchy problem for hyperbolic conservation laws with a relaxation term*, Proc. Royal Soc. Edinburgh, **126A** (1996), 821-828.

[KL2] C. Klingenberg and Y.-G. Lu, *Existence of solutions to hyperbolic conservation laws with a source*, Commun. Math. Phys., **187** (1997), 327-340.

[KL3] C. Klingenberg and Y.-G. Lu, *The vacuum case in Diperna's paper*, J. Math. Anal. Appl., **1225** (1998),679-684.

[Kr] S. N. Kruzkov, *First order quasilinear equations in several independent variables*, Mat. SB. (N.S.), **81** (1970), 228-255.

[LSU] O. A. Ladyzhenskaya, V. A. Solonnikov and N. N. Uraltseva, *Linear and quasilinear equations of parabolic type*, AMS Translations, Providence, 1968.

[LM] C. Lattanzio and P. Marcati, *The zero relaxation limit for the hydrodynamic Whitham traffic flow model*, J. Diff. Eqs., **141** (1997), 150-178.

[La1] P. D. Lax, *Hyperbolic systems of conservation laws II*, Comm. Pure Appl. Math., **10** (1957), 537-566.

[La2] P. D. Lax, *Shock waves and entropy*. In: Contributions to Nonlinear Functional Analysis, edited by E. Zarantonello, Academia Press: New York, 1971, 603-634.

[La3] P. D. Lax, *Hyperbolic Systems of Conservation Laws and the Mathematical Theory of Shock Waves*, SIAM, Philadelphia, 1973.

[Le] A. Y. LeRoux, *Numerical stability for some equations of gas dynamics*, Mathematics of Computation, **37** (1981), 435-446.

[LLL] R. J. Leveque, P. B. van Leer and H. C. Lee, *Model systems for reacting flow*, Final Report, NASA-Ames University Consortium NCA2-188, 1989.

[Lin] P.-X. Lin, *Young measures and an application of compensated compactness to one-dimensional nonlinear elastodynamics*, Trans. Am. Math. Soc., **329** (1992), 377-413.

[LPS] P. L. Lions, B. Perthame and P. E. Souganidis, *Existence and stability of entropy solutions for the hyperbolic systems of isentropic gas dynamics in Eulerian and Lagrangian coordinates*, Comm. Pure Appl. Math., **49** (1996), 599-638.

[LPT] P. L. Lions, B. Perthame and E. Tadmor, *Kinetic formulation of the isentropic gas dynamics and p-system*, Commun. Math. Phys., **163** (1994), 415-431.

[Liu] T.-P. Liu, *Hyperbolic conservation laws with relaxation*, Commun. Math. Phys., **108** (1987), 153-175.

[LW] T.-P. Liu and J.-H. Wang, *On a hyperbolic system of conservation laws which is not strictly hyperbolic*, J. Diff. Eqs., **57** (1985), 1-14.

[LWu] I.-S. Liu and Y.-M. Wu, *Vanishing relaxation limit of viscoelasticity*, Math. Mech. Solids, **1** (1996), 227-241.

[Lu1] Y.-G. Lu, *Convergence of solutions to nonlinear dispersive equations without convexity conditions*, Appl. Anal., **31** (1989), 239-246.

[Lu2] Y.-G. Lu, *Convergence of the viscosity method for nonstrictly hyperbolic conservation laws*, Commun. Math. Phys., **150** (1992), 59-64.

[Lu3] Y.-G. Lu, *Cauchy problem for an extended model of combustion*, Proc. Royal Soc. Edinburgh, **120A** (1992), 349-360.

[Lu4] Y.-G. Lu, *Convergence of the viscosity method for a nonstrictly hyperbolic system*, Acta Math. Sci., **12** (1992), 230-239.

[Lu5] Y.-G. Lu, *Global Hölder continuous solution of isentropic gas dynamics*, Proc. Royal Soc. Edinburgh, **123A** (1993), 231-238.

[Lu6] Y.-G. Lu, *Convergence of the viscosity method for some nonlinear hyperbolic systems*, Proc. Royal Soc. Edinburgh, **124A** (1994), 341-352.

[Lu7] Y.-G. Lu, *Cauchy problem for a hyperbolic model*, Nonlinear Anal., TMA, **23** (1994), 1135-1144.

[Lu8] Y.-G. Lu, *Convergence of viscosity solutions to a nonstrictly hyperbolic system*, in Advances in Nonlinear Differential Equations and Related Areas, G. Chen, Y., Li and X. Zhu, D, Cao, eds., World Scientific, (1998), 250-266.

[Lu9] Y.-G. Lu, *Singular limits of stiff relaxation and dominant diffusion for nonlinear systems*, J. Diff. Eqs., **179** (2002), 687-713.

[Lu10] Y.-G. Lu, *Singular limits of stiff relaxation for a nonstrictly hyperbolic system*, to appear.

[Lu11] Y.-G. Lu, *Global weak solution for a symmetric, hyperbolic system*, to appear.

[Lu12] Y.-G. Lu, *Viscosity and relaxation approximations for mixed type equations modeling reacting flows*, to appear.

[LK1] Y.-G. Lu and C. Klingenberg, *The Cauchy problem for hyperbolic conservation laws with three equations*, J. Math. Anal. Appl., **202** (1996), 206-216.

[LK2] Y.-G. Lu and C. Klingenberg, *The relaxation limit for systems of Broadwell type*, Diff. Int. Eqs., **14** (2001), 117-127.

[LMR] Y.-G. Lu, I. Mantilla and L. Rendon, *Convergence of approximated solutions to a nonstrictly hyperbolic system*, Advanced Nonlinear Studies, **1** (2001), 65-79.

[LuW] Y.-G. Lu and J.-H. Wang, *The interactions of elementary waves of nonstrictly hyperbolic system*, J. Math. Anal. Appl., **166** (1992), 136-169.

[Ma] A. Majda, *A qualitative model for dynamic combustion*, SIAM J. Appl. Math., **41** (1981), 70-93.

[Mu] F. Murat, *Compacité par compensation*, Ann. Scuola Norm. Sup. Pisa, **5** (1978), 489-507.

[Na] R. Natalini, *Convergence to equilibrium for the relaxation approximations of conservation laws*, Comm. Pure Appl. Math., **49** (1996), 795-823.

[Ni] T. Nishida, *Global solution for an initial-boundary-value problem of a quasilinear hyperbolic system*, Proc. Jap. Acad., **44** (1968), 642-646.

[NS] T. Nishida and J. Smoller, *Solutions in the large for some nonlinear hyperbolic conservation laws*, Comm. Pure Appl. Math., **26** (1973), 183-200.

[Oe1] K. Oelschläger, *On the connection between Hamiltonian many-particle systems and the hydrodynamical equation*, Arch. Rat. Mech. Anal., **115** (1991), 297-310.

[Oe2] K. Oelschläger, *An integro-differential equation modelling a Newtonian dynamics and its scaling limit*, Arch. Rat. Mech. Anal., **137** (1997), 99-134.

[Ol] O. Oleinik, *Discontinuous solutions of nonlinear differential equations*, English transl. in Am. Math. Soc. Transl Ser., **26** (1957), 95-172.

[Pe] B. Perthame, *Kinetic Formulations*, Oxford Univ. Press, 2002.

[Pl] T. Platowski and R. Illner, *Discrete models of the Boltzmann equation: a survey on the mathematical aspects of the theory*, SIAM Rev., **30** (1988), 213-255.

[RAA1] H. K. Rhee, R. Aris and N. R. Amundsen, *On the theory of multicomponent chromatography*, Phil. Trans. Royal. Soc. London, **267A** (1970), 419-455.

[RAA2] H. K. Rhee, R. Aris and N. R. Amundsen, *First Order Partial Differential Equations*, Vols. I-II, New York: Prentice Hall, 1986, 1989.

[RSK] J. Rubinstein, P. Sternberg and B. Keller, *Fast reaction, slow diffusion, and curve shortening*, SIAM J. Appl. Math., **49** (1989), 116-133.

[Sc] S. Schochet, *The instant-response limit in Whitham's nonlinear traffic-flow model: Uniform well-poseness and global existence*, Asymptotic Analysis, **1** (1988), 263-282.

[SC] M. E. Schonbek, *Convergence of solutions to nonlinear dispersive equations*, Comm. Partial Diff. Eqs., **7** (1982), 959-1000.

[Se1] D. Serre, *La compacite par compensation pour les systemes hyperboliques non lineaires de deux equations a une dimension despace*, J. Math. Pure Appl., **65** (1986), 423-468.

[Se2] D. Serre, *Solutions à variations bornées pour certains systèmes hyperboliques de lois de conservation*, J. Diff. Eqs., **68** (1987), 137-168.

[Se3] D. Serre, *Systems of Conservation Laws, I: Hyperbolicity, entropies, shock waves*, translated by I. N. Sneddon, Cambridge University Press, 1999.

[Se4] D. Serre, *Systems of Conservation Laws, II: Geometric structures, oscillations, and initial-boundary value problems*, translated by I. N. Sneddon, Cambridge University Press, 2000.

[Sh] J. Shearer, *Global existence and compactness in L^p for the quasilinear wave equation*, Comm. Partial Diff. Eqs., **19** (1994), 1829-1877.

[SSMP] M. Shearer, D. G. Schaeffer, D. Marchesin and P. J. Paesleme, *Solution of the Riemann problem for a prototype 2×2 system of nonstrictly hyperbolic conservation laws*, Arch. Rat. Mech. Anal. **97** (1987), 299-320.

[Si] C. G. Simader, *On Dirichlet's boundary value problem*, Lecture Notes in Math., **268**, Springer-Verlag, Berlin-Heidelberg-New York, 1972.

[Sm] J. Smoller, *Shock Waves and Reaction-Diffusion Equations*, Springer-Verlag, Berlin-Heidelberg-New York, 1983.

[So] S. Sobolev, *Partial Differential Equations of Mathematical Physics*, Pergamon Press, Oxford, 1964.

[St] J. J. Stoker, *Water Waves*, Interscience, New York, 1957.

[Sz] A. Szepessy, *An existence result for scalar conservation laws using measure valued solutions*, Commun. Partial Diff. Eqs., **14** (1989), 1329-1350.

[Ta] T. Tartar, *Compensated compactness and applications to partial differential equations*, In: Research Notes in Mathematics, Nonlinear Analysis and Mechanics, Heriot-Watt symposium, Vol. **4**, ed. R. J. Knops, Pitman Press, London, (1979).

[Te] B. Temple, *Systems of conservation laws with invariant submanifolds*, Trans. of Am. Math. Soc., **280** (1983), 781-795.

[Tr] H. Triebel, *Interpolation Theory, Function Spaces, Differential Operators*, North-Holland, Amsterdam-New York-Oxford-Tokyo, 1978.

[TW] A. Tveito and R. Winther, *On the rate of convergence to equilibrium for a system of conservation laws including a relaxation term*, SIAM J. Math. Anal., **28** (1997), 136-161.

[Tz] A. Tzavaras, *Materials with internal variables and relaxation to conservation laws*, Arch. Rat. Mech. Anal., **146** (1999), 2, 129-155.

[Wh] G. B. Whitham, *Linear and Nonlinear Waves*, John Wiley and Sons, New York, 1973.

[Yo] K. Yosida, *Functional Analysis*, Springer, New York, 1968.

Index